The struggle for market power

During the Industrial Revolution, class relations were defined largely through the struggle to control the terms of exchange in the market. Integrating aspects of economic and social history as well as industrial sociology, this book examines the sources of the perception of the market on the part of both capital and labor and the elaboration of their alternative market ideologies. Of particular import is the argument that working-class culture expressed a fundamental acceptance of the utility of the market, a point that is supported by a detailed analysis of the labor process, workplace bargaining, and early-nineteenth-century trade unionism. The determination of market relations in this era therefore became a function of both class power and ideological prescription.

The struggle for market power

Industrial relations in the British coal industry, 1800–1840

JAMES A. JAFFE
University of Wisconsin

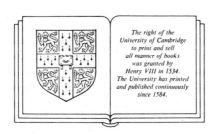

The right of the
University of Cambridge
to print and sell
all manner of books
was granted by
Henry VIII in 1534.
The University has printed
and published continuously
since 1584.

CAMBRIDGE UNIVERSITY PRESS

Cambridge

New York   Port Chester   Melbourne   Sydney

Published by the Press Syndicate of the University of Cambridge
The Pitt Building, Trumpington Street, Cambridge CB2 1RP
40 West 20th Street, New York, NY 10011, USA
10 Stamford Road, Oakleigh, Melborne 3166, Australia

© Cambridge University Press 1991

First published 1991

Printed in the United States of America

Library of Congress Cataloging-in-Publication Data
Jaffe, James Alan, 1954–
The struggle for market power : industrial relations in the
British coal industry, 1800–1840 / James A. Jaffe.
p.   cm.
Includes bibliographical references.
Includes index.
ISBN 0-521-39146-6
1. Collective bargaining – Coal mining industry – England –
History – 19th century.   2. Wages – Coal miners – England –
History – 19th century.   3. Piece-work – England – History – 19th
century.   I. Title.
HD6668.M615J34   1991
331'.0422334'09427 – dc20   90-43069
                                                                CIP

British Library Cataloguing in Publication Data
Jaffe, James A.
The struggle for market power : industrial relations in
the British coal industry, 1800–1840.
1. Great Britain. Coal industries. Industrial relations,
history
I. Title
331.04223340941

ISBN 0-521-39146-6 hardback

For Deborah

Contents

Contents

Acknowledgments

This book has taken many years to complete. It began as a dissertation submitted toward the completion of a Ph.D. at Columbia University in 1984. In the course of the original and subsequent research, writing, and rewriting, I was aided, encouraged, cajoled, and influenced by many friends. Among them, Isser Woloch not only taught me social history as a graduate student but has continued to help me as both confidante and critic. The late Stephen Koss guided this book through its early stages as a thesis. Naturally, I have missed his presence and influence, and I hope this book still bears his imprint.

Several institutions helped fund portions of this work, and I would like to thank them. Columbia University funded the first stages; the National Endowment for the Humanities supported work on Lord Londonderry's estates; and both the University of Wisconsin, Whitewater, and the University of Wisconsin system provided a year's leave during which this book was finally completed. Many friends took the time to read and comment on portions of the book along the way: James Donnelly, Joseph Donner, Geoffrey Field, Michael Hanagan, William Hausman, Lawrence Klein, William Maynard, Bob Montera, and Donald Reid. The anonymous readers for Cambridge University Press provided particularly stimulating and helpful criticisms. Earlier versions of portions of the book appeared in the *Journal of British Studies, Journal of Social History,* and *Northern History.* I thank the editors of those journals for permission to reproduce sections of them here.

In England, the help and guidance of many archivists, librarians, and staffs were unstinting. In Durham, the services of many people were very useful, but Mr. J. F. Hargrave's assistance at the Record Office, as well as his kindnesses outside of it, are especially appreciated. At the Northumberland Record Office, the former county archivist R. M. Gard provided aid in securing materials from the North of England Institute of Mining and Mechanical Engineers that was of incalculable value. Miss Annette Burton also helped me in locating material necessary for the completion of this project. I owe special thanks to Lord Lambton for permission to use material from the Lambton Mss. at the Lambton Estate Office. The archivists and librarians at the Tyne and Wear County Archives Department, the North of England Institute of Mining and Mechanical Engineers, the Newcastle Central Library, and the Department of Palaeography and Diplomatic at the University of Durham all helped me at various stages of this project. Needless to say, the work of all these persons is gratefully acknowledged.

I owe a special debt to my family. My parents have been supportive of my endeavors, and this has always been a great source of strength. Finally, this book is dedicated to Deborah, whose confidence, faith, and love sustained me through the long years of research and writing. Her influence appears on every page. Our daughter, Olwen, will surely neither remember nor quite comprehend how much she has contributed to this book.

Abbreviations

D.C.R.O.	Durham County Record Office
H.O.	Home Office
NCB	National Coal Board
N.C.L.	Newcastle Central Library
N.E.I.M.M.E.	North of England Institute of Mining and Mechanical Engineers
N.R.O.	Northumberland Record Office
P.P.	Parliamentary Papers
P.R.O.	Public Record Office
T.W.C.A.	Tyne and Wear County Archives

Introduction

This book is about the struggle to define the market. It is based upon the premise that the terms of exchange between classes are historically constructed through struggles over the control of market relations. These struggles, rather than those over the control of the labor process, formed the substance of industrial relations in the northern coal trade of industrializing England.

Underlying this premise is the corollary that early industrial working-class culture expressed a fundamental acceptance of the utility of the market. The source of this acceptance was not, as others have argued or implied, workers' experience in the market for food or consumer goods.[1] Instead, market relations penetrated local society largely through experience gained in the workplace, particularly the practice of bargaining over piece rates.

Historical analysis of the social impact of the wage form is surprisingly lacking. In the British economy as a whole, it is quite clear that piece rates became increasingly more popular among employers as the nineteenth century progressed.[2] However, it is by no means as clear whether, or the extent to which, the adoption of the piece-rate system of payments involved a concomitant adoption of wage bargaining, collective or otherwise, or forms of unionization. It may well be, as Carter Goodrich suggested, that in many trades the implementation of piece-rate payments and the growth of trades unions were connected.[3] It is certainly striking that boards of conciliation

---

[1] See, e.g., Dale Edward Williams, "Morals, Markets and the English Crowd in 1766," *Past & Present*, no. 104 (August 1984), 56–73, and on the "consumer revolution" of the eighteenth century, see Neil McKendrick, John Brewer, and J. H. Plumb (eds.), *The Birth of a Consumer Society: The Commercialization of Eighteenth-Century England* (Bloomington: Indiana University Press, 1982).
[2] John Rule, *The Labouring Classes in Early Industrial England, 1750–1850* (London: Longman Group, 1986), 120–6.
[3] Carter Goodrich, *The Frontier Of Control* (1920, repr. London: Pluto Press, 1975), 165, suggested that industries operating on piece-rate systems "naturally show more instances of union activity."

and arbitration were pioneered, and became most successful, in trades such as hosiery and footwear in which piece work was most prevalent.[4] Similarly, in textiles, the negotiated settlement of price lists after 1850 for the Bolton spinners and Blackburn weavers testifies to the connection between piece work and collective bargaining.[5] Indeed, a culture of bargaining also may have been a facet of the textile workers' community. Thus, a witness before Sadler's Committee in 1832 testified that "I always bargained for wages ever since I began to work for wages."[6]

The problem of identifying the social impact of the wage form and bargaining is compounded by the existence of a variety of forms of wage bargaining in the nineteenth century: some between individuals and employers, others between work groups and employers, still others over the establishment of trade price lists.[7] In trades such as bespoke tailoring and flint glassmaking, for example, hourly wages served to mask piece rates that were construed in the form of a "log" or a "move."[8] In such cases, bargaining was still an important aspect of industrial relations, but was more likely to

---

[4] H. A. Clegg, Alan Fox, and A. F. Thompson, *A History of British Trade Unions Since 1889*, Volume 1: 1889–1910 (Oxford: Clarendon, 1964), 24–6; William Lazonick, "Industrial Relations and Technical Change: The Case of the Self-acting Mule," *Cambridge Journal of Economics* 3 (1979), 246–7, 257–8. The latter, however, emphasizes that the adoption of collective bargaining procedures occurred largely as an accommodative response among employers who felt themselves caught in a highly competitive industry rather than as rooted in work itself.

[5] The Webbs themselves noted the "interesting parallelism" between the forms of organization and objectives of the coal miners and the cotton operatives. They explained the similarity, however, not by the nature of work and industrial relations but, predictably, by the special position of trade union officials and the unionists adherence to traditional living standards. Sidney Webb and Beatrice Webb, *The History of Trade Unionism*, new ed. (London: Longman Group, 1911) 298–9.

[6] Quoted in Robert Gray, "The Languages of Factory Reform in Britain, c. 1830–1860," in Patrick Joyce (ed.), *The Historical Meanings of Work* (Cambridge University Press, 1987), 150. The work of Michael Huberman outlines a contractual model of labor relations; however, the actual bargaining process in the textile industry has yet to be fully adumbrated. See Michael Huberman, "The Economic Origins of Paternalism: Lancashire Cotton Spinning in the First Half of the Nineteenth Century," *Social History*, 12, no. 2 (1987), 177–92. See also, Mary Rose, Peter Taylor, and Michael Winstanley, "The Economic Origins of Paternalism: Some Objections," *Social History*, 14, no. 1 (1989), 89–98; Michael Huberman, "The Economic Origins of Paternalism: Reply to Rose, Taylor, and Winstanley," ibid., 99–103.

[7] Goodrich, *Frontier of Control*, 165–6; on district price lists in the cotton industry, see William Lazonick, "The Cotton Industry," in Bernard Elbaum and William Lazonick (eds.), *The Decline of the British Economy* (Oxford: Clarendon, 1986), 24–8.

[8] Takao Matsumura, *The Labour Aristocracy Revisited: The Victorian Flint Glass Makers, 1850–1880* (Manchester: Manchester University Press, 1983), 48–51; D. F. Schloss, *Methods of Industrial Remuneration*, 3rd ed. (London: Williams & Norgate 1898), 26–7, also cited in Matsumura, *Labour Aristocracy Revisited*, 75.

Introduction

have involved the establishment of quantities or production quotas rather than piece rates per se.[9]

John Rule's work on the "tribute system" among Cornish miners presented another variation in which miners bargained to work certain portions of the mine with their payment based on a percentage of the value of ore delivered aboveground.[10] (Unlike the northern coal miners, however, the Cornish miners bore the full burden of changes in the geology of ore veins that could lead equally to prosperity or poverty.) In the pottery industry of the early twentieth century, Richard Whipp has described a still different set of industrial relations premised upon bargaining that was unevenly spread across firms, work groups, and subindustries. Nonetheless, in Whipp's estimation, "localized, small-scale bargaining was the most common form in which pottery workers experienced industrial relations."[11]

Therefore, while a good deal of information exists about the forms and structures of industrial remuneration, there is still much work to be done on the extent to which and why these systems of wage bargaining were accepted; on how the ethic of bargaining was internalized, as it seems to have been among textile operatives, pottery workers, and coal miners; and, whether, as some have argued, the adoption of piece rates invariably led to alienation, the intensification of labor, and the loss of control.[12] On this last point, Schloss noted cases in which the introduction of time work was resisted by workers when trades were mechanized. In those instances, workers' defense of piece work was viewed as a way to share in the growth of productivity and to avoid further exploitation.[13]

---

[9] E. A. Pratt, *Trade Unionism and British Industry* (London: Murray, 1904), 97, cited in Matsumura, *Labour Aristocracy Revisited*, 75, n.3.

[10] Rule, *Labouring Classes*, 124–5; idem, "The Labouring Miner in Cornwall: A Study in Social History" (Ph.D. dissertation, University of Warwick, 1971).

[11] Richard Whipp, "Work and Social Consciousness: The British Potters in the Early Twentieth Century," *Past & Present*, no. 119 (May 1988), 147.

[12] The ambivalence of piece rates is certainly an extremely important point for the history of labor and the labor movement. Schloss, of course, noted this conflict, and labor historians tend to divide along similar lines. See Rule, *Labouring Classes*, 120–6, and K. McClelland and A. Reid, "Wood, Iron and Steel" in Royden Harrison and Jonathan Zeitlin (eds.), *Divisions of Labour Skilled Workers and Technological Change in Nineteenth Century England* (Sussex: Harvester, 1985) 164, for arguments that piece rates wrested control from workers and aggravated alienation. However, piece rates as Goodrich noted, were sometimes defended by labor precisely because they conferred a greater degree of job control and independence. See Goodrich, *Frontier of Control*, 163–4. Richard Price has similarly noted that piece work's social impact was inherently ambiguous. See Richard Price, "Labour Process and Labour History," *Social History* 8, no. 63, (1983) and idem, "Rethinking Labour History: The Importance of Work," in James E. Cronin and Jonathan Schneer (eds.), *Social Conflict and the Political Order in Modern Britain* (New Brunswick, N.J.: Rutgers University Press, 1982), 201–2.

[13] Schloss, *Methods of Industrial Remuneration*, 56–9. Similarly, Lazonick notes the case of the

In a still broader sense, the study of piece work and its relation to forms of bargaining can be linked to the structural development of the British economy. As Elbaum and Lazonick suggest, the institutionalization of piece work and bargaining contributed to the rigidities of British industry in the late nineteenth and early twentieth centuries and inhibited competitive responses to changing markets.[14] In the case of Lancashire textiles, Lazonick has argued persuasively that the adoption of best-practice techniques was constrained in part by the conciliatory structure of industrial relations that were drawn against a background of cutthroat domestic and international competition.[15]

Part of the argument this book puts forward, therefore, is that bargaining was accepted as the principal terrain of industrial relations, and its ubiquity throughout the coal industry negotiated relations between classes. Through bargaining over piece rates, the miners of the Northeast came to understand and accept market relations. In one sense, therefore, the work presented here corresponds to Richard Price's recent survey of British labor history in which it is argued that social and industrial relations during the Industrial Revolution were shaped principally by the impact of the market rather than the restructuring of production relations.[16] However, unlike the evidence presented by Price, the coal miners' acceptance of market relations precluded a resistance to exploitation based upon notions of customary or lost independence.[17] Instead, the acceptance of market relations necessarily entailed an implicit rejection of the dichotomy, which is often an anachronistic one, employed by social historians between the customary, nonmarket culture of the preindustrial and early industrial working class and the hegemonically imposed market culture of competitive individualism under industrial capitalism. The northern miners of the early nineteenth century exhibited a culture and an ideology both aware of and engaged with the market. Indeed, to them, the market was their industrial culture.

Equally important, however, is the argument that the acceptance of market principles did not entail the acceptance of bourgeois political economy. It was readily apparent to workers with experience in wage bargaining that market mechanisms were not objectively determined; the hand was not

Lancashire minders whose negotiated wage lists "ensured them a share (and usually a significant one) in productivity gains" achieved by running self-acting mules faster. See Lazonick, "Industrial Relations and Technical Change," 256.
[14] Lazonick, "Cotton Industry," 27; Bernard Elbaum, "The Steel Industry before World War I," in Elbaum and Lazonick (eds.), *Decline of the British Economy*, 69–71.
[15] Lazonick, "Industrial Relations and Technical Change," 258.
[16] Richard Price, *Labour in British Society: An Interpretative History* (1986; repr. London: Routledge, 1990), 20–48.
[17] Ibid., 45.

invisible. Certainly in the coal industry, as elsewhere, where masters sought to control and monopolize markets, it is inappropriate to speak of an ideological conflict between artisinal "regulation" and industrial laissez-faire.[18] Both industry and labor sought to turn the terms of the market to their own best advantage. Industrial relations, therefore, were the result of struggles and compromises over the definition and control of several different markets, particularly the disposition of the product and labor markets. While still accepting the market's ultimate rationality, both the volume of production and the structure of the labor market were understood to be subject to the organizing power of both capital and labor. Logically, the determination of the precise character of these market relations in any historical epoch ultimately rested upon the social and economic power each class could bring to the struggle. In sum, the structuring of market relations was a problem of both power and prescription.

The market, therefore, like E. P. Thompson's famous definition of class, must be viewed in historical terms, that is, as a process and a relationship, not a structure or a thing.[19] Objections to the reification of the market and the erection of "laws" of supply and demand, of course, underlay Marx's attacks on political economy. It is the purpose of this book to examine the historical construction of the market and its characteristic forms of industrial and social relations during the era of the Industrial Revolution.

To argue thus is to accept the importance of the experience of class, albeit from a different angle than previous work in the field. To some labor historians or industrial sociologists, this point may appear to be superfluous. However, the history of labor is now being enriched by what one historian has called the "social history of language."[20] And to those whose primary interest is in the "autonomy" of language or the primacy of "discourse," this work may seem insufficiently abstract and rooted too firmly in a materialist conception of history. Still, while the study of language can become an essential component of the history of the working class – indeed, as I have said, the problem of the market was in part a problem of prescription – it is contended here that the identification and definition of class interests manifested themselves principally through the labor process and its penumbra of industrial relations.

To reconstruct the struggle for power in the market, industrial relations needed to be defined in the broadest sense. This is so particularly because

---

[18] Ibid., 29.

[19] E. P. Thompson, *The Making of the English Working Class* (New York: Vintage, 1963), 9–10.

[20] Gregory Claeys, *Citizens and Saints: Politics and Anti-politics in Early British Socialism* (Cambridge University Press, 1989), 18; Gareth Stedman Jones, "Rethinking Chartism," in Stedman Jones, *Languages of Class: Studies in English Working Class History, 1832–1982* (Cambridge University Press 1983).

the struggle was not restricted solely to the shopfloor. Thus, in this book, not only are the goals, tactics, and methods of shopfloor relations and the organization of production examined, but equal weight is given to the analysis of the impact of the market on family and community, industrial paternalism, the finance and structure of industry, and religious and political ideologies.

Given this wide field of inquiry, it is natural that many fields of historical controversy have been traversed. This has made the problem of organization a complex one. The book, therefore, generally eschews a chronological framework in favor of a thematic organization. The first three chapters analyze the stucture and ideology of industry. Chapters 4, 5, and 6 discuss the expression of the market in working-class life, labor, and ideology. The succeeding two chapters are a detailed reconstruction of Tommy Hepburn's union movement of 1831–2. In these final chapters, given one of the common tools of the social historian, the analysis of a strike, it can be shown that the struggle to control the market shaped the context and character of class relations in the early industrial era.

1

Capital and credit

"*Les dés sont pipés*," Marx wrote in the first volume of *Capital*.[1] He was referring to the fact that the forces of supply and demand failed to operate objectively in the labor market. Instead, the "laws" of political economy were structured to suit the needs of capital: Demand for labor was mitigated by mechanization, which in turn increased the supply of labor. "It is not a case of two independent forces working on each other," Marx concluded. "Capital acts on both sides at once."[2] Conversely, when workers recognized the inequity of the "free" labor market and sought to create trade unions, their employers "cry out at the infringement of the 'eternal' and so to speak 'sacred' law of supply and demand."[3]

Therefore, the dice are loaded. Yet perhaps it would be more proper to say that the dice become loaded. The terms of market relations are historically contested; there is not one law of supply and demand but several competing prescriptive understandings of such a law. The trade unionists of the early nineteenth century, for example, generally did not seek to abrogate the operation of market forces. However, they did seek to adapt and define them according to their needs. This, it might be claimed, is what separated trade unionism and cooperation from socialism. Similarly, the convenient support of a free labor market that Marx noted among capitalists did not apply equally to the market for their commodities. One need only recall Adam Smith's oft-quoted phrase: "People of the same trade seldom meet together, even for merriment and diversion, but the conversation ends in a conspiracy against the public, or in some contrivance to raise prices."[4]

The laws of supply and demand, therefore, are subject to both interpretation and manipulation. Their formulation is at any one time the result of a complex interplay of social, economic, and political forces. Yet the defi-

---

[1] Karl Marx, *Capital: A Critique of Political Economy*, vol. 1 (New York: Vintage, 1977), 793.
[2] Ibid.
[3] Ibid.
[4] Adam Smith, *An Inquiry into the Nature and Causes of the Wealth of Nations*, Edwin Cannan, ed. (Chicago: University of Chicago Press, 1976), 144.

nition of market relations both for each class and between classes does not take place solely in the labor market, as one might infer from Marx's example. The disposition of credit or financial arrangements, commodity markets, and labor markets contributes to the struggle to determine the definition of the operation of the market.

Defining these market relations is an exercise in the articulation of an ideology as well as the realization of tangible goals. It is thus a political process, the resolution of which is ultimately based on the organization and deployment of class power. The trade union movement of early-nineteenth-century England thus must be seen as an attempt to capture the terms upon which the marketplace in civil society was being restructured. It was this struggle over the definition and structuring of market relations, rather than that over production relations generally or the labor process in particular, that formed the essential element of the reconstitution of class relationships during the Industrial Revolution.

The process of industrialization inevitably involved a redefinition of market relations. However, the character of that process has itself taken on a different perspective. Once a seamless web of the story of human triumph over nature, the history of British industrialization is now faced with complex problems of timing, scale, and context. N. F. R. Crafts has revealed an Industrial Revolution marked by slow growth, low productivity, and low-wage, unskilled labor. The same achievement that David Landes once hailed as the triumph of "the quality of inputs" over the traditional material factors of production[5] Crafts now characterizes as an economic and social commitment to industries that were tied to a low level of return on investments, that failed to accumulate human capital, and that lagged behind later industrializers in research and development.[6]

As Clapham's "industrial revolution in slow motion" has taken hold once again, it is perhaps not surprising that interest in handicraft production has witnessed a marked resurgence.[7] Generated in part by Franklin Mendel's influential theory of protoindustrialization, historians have begun not only to think anew of the timing and character of industrialization, but also to

---

[5] David Landes, *The Unbound Prometheus, Technological Change and Industrial Development in Western Europe from 1750 to the Present* (Cambridge University Press, 1969), 79–80.
[6] N. F. R. Crafts, *British Economic Growth during the Industrial Revolution* (Oxford: Clarendon, 1985).
[7] See, e.g., Maxine Berg, *The Age of Manufactures, 1700–1820* (Fontana, 1985), and Maxine Berg, Pat Hudson, and Michael Sonenscher (eds.), *Manufacture in Town and Country before the Factory* (Cambridge University Press, 1983).

consider the dynamics of the family, the role of the division of labor, and the impact of custom on economic development.[8]

The discovery of a new frontier like protoindustrialization is heady stuff for the historian, but when we come to consider the coal industry during industrialization, many of these new conceptual tools cannot be made to fit.[9] The dominant theme of protoindustrialization, for example, that rural hand-icraft production for the market was a distinct historical phase preceding industrialization, bears little relation to the history of coal production in the Northeast during the eighteenth century.[10] As Michael Flinn pointed out, particularly in the Northeast, the British coal industry during the era of handicraft production was not "pre-industrial"; it was characterized by large, centralized units of production, the application of sophisticated technology, and heavy fixed capital investments. In this sense, the coal industry may not have paralleled the experience of other protoindustrial sectors of the British economy, but it did project the image of their future.[11]

Indeed, the capital demands of firms in the coal trade, both fixed and circulating, were so extensive that all manner of financing, including plowed-back profits, merchant credit, mortgaging, and cooperative purchasing, was employed. However, these demands could also act as restraints upon both industrial development and labor relations. The availability of resources for both fixed and circulating capital was a crucial determinant not only of the ultimate success or failure of the firm but also of its ability to deal both with erstwhile competitors and labor. Thus, the demands, sources, and availability of capital outlined the context of both action and ideology for industrialists in the coal trade.

Yet whether in terms of Marxian primitive accumulation or in the language of protoindustrialization, preindustrial characteristics of production are not without their relevance to the history of the coal industry and its industrial

---

[8] L. A. Clarkson, *Proto-Industrialization: The First Phase of Industrialization?* (London: Macmillan, 1985).

[9] D. C. Coleman has argued that the economic development of the Northeast and South Wales belie the significance of protoindustrialization for the British economy. See D. C. Coleman, "Proto-industrialization: A Concept Too Many," *Economic History Review*, 2nd ser., 36, no. 3 (1983); Clarkson, *Proto-Industrialization*, 54–5; and, Neil Evans, "Two paths to economic development: Wales and the North-east of England," in Pat Hudson, ed., *Regions and Industries: A Perspective on the Industrial Revolution in Britain* (Cambridge University Press, 1989), 203. The problem is also mentioned in passing in idem, *The Genesis of Industrial Capital: A Study of the West Riding Wool Textile Industry c. 1750–1850* (Cambridge University Press, 1986), 59.

[10] Clarkson, *Proto-Industrialization*, 15–17; Berg, *Age of Manufactures*, 77–83.

[11] Michael W. Flinn, *The History of the British Coal Industry, 1700–1830: The Industrial Revolution* (Oxford: Clarendon, 1984), 456–7.

relations. Market perceptions, industrial policy, and the logic of industrial relations in the coal industry were derived substantially from the perception of the industry as a trade and the coal miner as a commodity producer paid by the piece. Moreover, the relationship of land to industry and the agrarian origins of industrial capital, two important aspects of protoindustrial development, are also interpenetrating themes of great importance to the history of the northern coal industry.

Fixed capital and working costs

During the period of industrialization, landownership and mining were closely connected. Mining profits and mineral rents formed a substantial portion of the incomes of many landed families.[12] For some, investments in mining may have originally been facilitated simply by the fortuitous location of their estates over coal deposits. However, by the eighteenth and nineteenth centuries, land purchases and leasings by coal owners were being undertaken as part of a systematic extension of enterprises in coal.[13]

The typical colliery of the nineteenth century was still closely intermeshed with its agricultural environs. Indeed, a colliery in the Durham and Northumberland coalfield was very much like a large, integrated agro-industrial combine. Whether operated by aristocrats, gentry, or nonnoble industrialists, coal production necessitated substantial agricultural inputs. This was so largely because of the need to feed and pasture colliery horses used to transport coal both above and below ground. The Londonderry estates, for example, maintained five colliery farms of nearly seven hundred total acres. The principal output of these farms was hay. In 1835, the farms produced 787 tons of hay, but they also produced small quantities of wheat, oats, beans, and turnips. In the same year, the colliery farms supported 104 horses valued at £545.[14]

Colliery farms also maintained stock of relatively significant value. In 1809, stocks of oat and hay at the Jarrow and Temple Main Colliery amounted to £1,147, or 4.6 percent of the total valuation of the collieries.[15] The five Londonderry farms produced feed and maintained stock valued at £4,755

---

[12] J. T. Ward, "Landowners and Mining," in J. T. Ward and R. G. Wilson, *Land and Industry: The Landed Estate and the Industrial Revolution* (New York: Barnes & Noble, 1971), 63–116.

[13] Michael Sill, "Landownership and Industry: The East Durham Coalfield in the Nineteenth Century," *Northern History*, 20 (1984), 146–66.

[14] D.C.R.O., Londonderry Papers, D/Lo/B 310(17), "Estimate of Value of Produce on Colliery Farms, 1835."

[15] Flinn, *History of the British Coal Industry*, 198.

in 1835, or just under 4 percent of the total valuation of the Rainton, Pensher, and Pittington collieries.[16]

The successful exploitation of a mine required that surface land rights for transport, housing, and farming be made as secure as possible in order to protect investments. In the 1830s and 1840s, as Michael Sill has shown, colliery companies spent several years leasing both underground and surface rights before the first sinkings for coal were attempted.[17] Certainly in the case of colliery farms, the supply of surface acreage seems to have posed less of a problem for the aristocratic coal owners, who were already owners of property, than for companies or partnerships in the Northeast. The two largest Londonderry colliery farms, Pensher (272 acres) and Grange Farm (306 acres), were both situated on freehold property, while the much smaller farms at Chilton Moor (57 acres) and West Rainton (33 acres) were sited on copyhold and leasehold land, respectively.[18] Colliery farms on the Londonderry estates, therefore, were concentrated on freehold land, where they were protected from the vagaries and added costs of cultivating leasehold property.

The value of this property necessarily amounted to substantial sums. The 527-acre Pensher estate, site of Pensher Colliery, was valued in fee (exclusive of coal reserves) at over £53,500 in 1834.[19] Obviously, the location of the colliery dramatically drove up the value of property, and the same can also be said for those lands possessed of wayleaves, that is, the negotiated right of coal transport over private property. Wayleaves were valued at twenty years' purchase price, so that a wayleave to the Stanhope & Tyne Railway worth £500 per annum over the Barmston estate in Durham raised the value in fee of that property by £10,000.[20] The fact that landed property could be leased, however, also meant that barriers to entry in this regard could be reduced significantly. Thus, initial investments could be reduced substantially where leases could be negotiated instead of the outright purchase of property.

The same, however, cannot be said for the capital embodied in winning and sinking a colliery and in the purchase of colliery machinery. Here, capital expenditures placed coal firms among the largest firms in the nation. Start-up costs (sinking the pits and winning, i.e., preparing, the colliery) for a

---

[16] D.C.R.O., Londonderry Papers, D/Lo/B 310(17); the date of the colliery valuation used is 1827: D/Lo/B 81 (16–18).
[17] Sill, "Landownership and Industry," 162–3.
[18] D.C.R.O., Londonderry Papers, D/Lo/E 503(3), "Valuation of Londonderry Estates in Durham, 1834."
[19] Ibid.
[20] Ibid.

relatively modest colliery on the Tyne amounted to over £25,000 in 1809.[21] And those collieries that followed the Hetton Coal Company into east Durham after 1822 required significantly greater preliminary investments. Monkwearmouth Colliery, for example, opened in 1826 after an initial investment of about £100,000. A presinking estimate of South Hetton Colliery's development was put at over £66,000, and Murton Colliery was opened in 1843 after five years of sinking at a cost estimated at between £250,000 and £300,000.[22]

Initial sinking and winning costs were supplemented over succeeding years by continuous investments to deepen and extend pits in order to maintain production.[23] In 1831, Matthias Dunn offered the following estimates for sinking one new pit at Hetton Colliery:[24]

Sinking 180 fathoms	£1800
6 Wedging Cribs	45
60 fathom tubbing	300
50 do. walling	100
Smithwork	106
[illegible] Plugmen, &c	250
Sundries unforeseen	199
Labour	£2800
Wedging Cribs	42
Tubbing £28	1680
Walling £5½	275
Horses Keep	360
Sinking Corves	20
Timber [£]68 Iron [£]191	259
Leather, &c [£]150	
House Rts. [£]114	264
Total	£5700

Maintaining production thus required substantial continuous investment: The Vane-Tempest collieries expended £14,878 on sinking new pits between 1815 and 1818, which then was followed by further capital outlays of approximately £46,000 to expand and improve the collieries under Lord Lon-

[21] N.E.I.M.M.E., Misc. Collections, 63/ZC/48/1, quoted in Flinn, *History*, 193.
[22] Robert L. Galloway, *A History of Coal Mining in Great Britain* (1882; repr. Newton Abbot: David & Charles, 1969), 203–10; Sill, "Landownership and Industry," 152; A. G. Kenwood, *Capital Formation in North East England, 1800–1913* (New York: Garland, 1985), 63–71.
[23] Flinn describes these as "running" investments in *History of the British Coal Industry*, 194–7.
[24] N.C.L., M. Dunn's Diary, 1 December 1831.

donderry's first years as life tenant between 1819 and 1824.[25] Similarly, the modernization of the Lambton collieries between 1823 and 1828, entailed expenditures of over £38,000.[26]

In terms of contemporary accounting practice, winning and sinking pits were often considered distinct from investments in machinery and related materials. The Vane-Tempest colliery accounts considered increases of stock as an "ordinary" expenditure, while the cost of sinking new pits was "extraordinary."[27] These accounting procedures make it extremely difficult to assess precisely the capital investments of northeastern mining firms. The most common sources of our knowledge are the surviving "valuations" that distinguish the value of "fixed stock" (machinery and equipment that could not be readily dismantled and removed, such as engines, staithes, and waggonways) from "moveable stock" (such as ropes, wagons, and coal stocks).[28] However, these accounts are themselves limited since there was no attempt to depreciate the stock nor were these accounts always based on the resale or market price of the equipment.[29]

Thus while it is exceedingly difficult to determine the exact degree of total investments in any one firm, one can get an indication of the scale of operations from the valuations of the colliery's assets. In this regard, the largest firms in the Northeast coal trade were among the largest firms in the country in the early nineteenth century. In 1827, the capital stock of the three Londonderry collieries was valued at over £120,000 (Table 1.1). Similarly, the capital stock of four Lambton (Durham) collieries in 1835 was valued at over £155,000 (Table 1.2). An estimate of the gross value of Hetton Colliery, the largest colliery in the northern coalfield by the 1830s, was put

[25] D.C.R.O., Londonderry Papers, D/Lo/B 41; R. W. Sturgess, *Aristocrat in Business: The Third Marquess of Londonderry as Coalowner and Portbuilder* (Durham County Local History Society, 1975), 36.

[26] Sturgess, *Aristocrat in Business*, 36; Lambton Mss., unnumbered, quoted in Flinn, *History of the British Coal Industry*, 194.

[27] D.C.R.O., Londonderry Papers, D/Lo/B 41.

[28] These definitions are generally at odds with those described as "dead" and "live" stock by Flinn, *History of the British Coal Industry*, 200, and are based partly on Buddle's valuation of the Londonderry collieries in 1827. See D.C.R.O., Londonderry Papers, D/Lo/B 81(16–18). In Buddle's valuation of the Lambton (Durham) collieries in 1835, the terms "dead" and "live" stock and "fixed" and "moveable" stock seem to be used interchangeably. See D.C.R.O., NCB I/JB/ 2087. However, in neither case do the valuations include the value of shafts, as Flinn contended, while steam engines appear to have been considered "fixed" or "dead", not "live" and "moveable" stock.

[29] Buddle, however, did indicate that fixed stock could be resold at about 40 percent below its valuation. D.C.R.O., NCB I/JB/2088; *Report from the Select Committee of the House of Lords appointed to take into Consideration the State of the Coal Trade in the United Kingdom* (hereafter *Select Committee of the House of Lords on the State of the Coal Trade*), P.P. (1830), vol. 8, 38.

Table 1.1. *Capital stock, Londonderry collieries, 1827*

Colliery	Fixed stock	Movable stock	Total
Rainton	£33,543	£37,609	£71,153
Penshaw	14,979	21,231	36,221
North Pittington	9,635	4,344	13,458
Total	57,635	63,184	120,822

Source: D.C.R.O., Londonderry Papers, D/Lo/B 81(16–18).

at £523, 297 in 1832.[30] In the case of the Lambton collieries, John Buddle estimated the total value of the works (apparently including coal reserves) at £271,140. This, notably, excluded the value of at least two dormant collieries, Cassop and Ludsworth, as well as the value of Sherburn Colliery (valued at over £80,000) for which no stock valuation exists. The total value of the Lambton works, including Sherburn, was estimated at £384,381.

Compared with total value, fixed and movable stock comprised 57.7 percent of the valuation of the earl of Durham's four active collieries. This proportion differs considerably from other early industrial firms in which fixed capital frequently made up only 10 to 20 percent of total valuation.[31] Moreover, the coal industry was unique in that the most significant portion of the remainder of capital was not tied up in circulating capital, as it was in textiles, for example, but in unexploited coal reserves that could be won and worked only at indeterminate costs.

While the validity and usefulness of valuations offers only a very rough guide to capital investments, some standardization of measurement may be achieved by calculating the ratio of capital to output. In Table 1.3, the reports of "capital embarked" by collieries is derived from a survey conducted by the industry in 1828. The chief claim to veracity for these figures is that they apparently formed the basis of John Buddle's widely accepted testimony to the House of Lords that estimated total Tyneside capital investment amounted to £1.5 million.[32] (The returns, as they were compiled, omit the figures for five collieries and are incomplete for two others.) While calculating total investments is fraught with difficulty, the same must also be said for any attempt to establish total output. Flinn has argued that the

---

[30] D.C.R.O., Londonderry Papers, D/Lo/C 142(883), Buddle to Londonderry, 7 September 1832.

[31] Phyllis Deane, *The First Industrial Revolution*, 2nd ed. (Cambridge University Press, 1979), 171. See also Landes, *Unbound Prometheus*, 75, and Sidney Pollard, "Fixed Capital in the Industrial Revolution in Britain," *Journal of Economic History*, 24: no. 3 (1964), 299–314.

[32] *Select Committee of the House of Lords on the State of the Coal Trade*, P.P. (1830), vol. 8, 34.

Table 1.2. *Capital stock, Lambton collieries, 1835*

Colliery	Fixed stock	Movable stock	Total
Cocken	£7,004	£2,073	£9,077
Lumley	19,222	4,125	23,347
Littletown	17,732	5,295	23,027
Newbottle	20,434	2,819	23,253
General stock[a]	57,197	20,455	77,652
Total	121,589	34,767	156,356

[a]General stock denotes equipment used jointly by more than one colliery, particularly staithes and railways.
Source: D.C.R.O., NCB I/JB/2087–8.

nominal figures of output collected by the northern coal trade underestimate total output by as much as 79.5 percent because foreign exports, waste, overweight sales, local landsales, miners' allotments, and colliery consumption were not included.[33] Taking these adjustments into consideration, Table 1.3 sets out the capital–output ratios for twenty-nine Tyneside collieries in 1828.

The significance of these numbers is that they reflect the substantial burden of capital costs that had accumulated in mining by the third decade of the nineteenth century. While Flinn has proved that the marginal productivity of capital in mining was rising into the nineteenth century and thus calculated a sharp drop in initial capital costs to £0.18 per ton in 1821–5 from £0.30 in 1803–14,[34] these proportions of total capital embarked indicate the cumulative effect of both initial and running investments on the capital structure of the firm. By the late 1820s, cumulative costs were in fact rising significantly, and financing of expansion and modernization was becoming increasingly more expensive relative to returns on output. Indeed, an event like the banking crisis of 1825–6 seriously threatened the finances of both the Lambton and Londonderry collieries, both of which had borrowed heavily to expand and modernize their facilities.[35] Furthermore, it is not surprising that coal companies such as the Hetton Coal Company should appear at around this time to compete with individuals of capital as mining required increasing capitalization. The 1820s thus marked a watershed in the northern

[33] Flinn, *History of the British Coal Industry*, 29–33.
[34] Ibid., 202.
[35] Lambton's debt was as high as £54,000 in 1826, according to Flinn, ibid., 208. For Londonderry, see the section in this chapter on cash and credit.

Table 1.3. *Capital–output ratios on Tyneside, 1828*

Colliery	Adjusted gross output (tons)	Total capital (£)	Fixed capital per ton (£)
Backworth	108,624	60,000	0.55
Burradon and Killingworth	156,529	120,000	0.77
Coxlodge	110,056	68,000	0.62
Fawdon	110,954	110,000	0.99
Heaton	108,116	44,500	0.41
Holywell	88,870	15,000	0.17
Hotspur	99,298	25,482	0.26
Percy Main	163,381	100,000	0.61
Wallsend	127,423	50,000	0.39
Willington	121,464	42,507	0.35
Walker	103,279	44,000	0.43
Wideopen	99,746	40,000	0.40
Benwell	65,663	30,000	0.47
Cramlington	69,301	40,000	0.58
Elswick	77,417	27,000	0.35
Felling	60,863	24,000	0.39
Heworth	77,016	35,000	0.45
Low Moor and South Moor	83,105	19,400	0.23
Manor Wallsend	92,970	39,000	0.42
Pelaw Main	81,477	35,000	0.43
Pontop and Garesfield	84,956	25,000	0.29
Seghill	90,863	51,120	0.56
Sheriff Hill	63,789	22,000	0.34
Springwell	77,573	40,000	0.52
Tanfield Moor	72,545	16,000	0.22
Team	95,858	35,000	0.37
Walbottle	71,945	50,000	0.69
Whitley	44,774	14,000	0.31
Wylam	61,728	30,000	0.49
Total	2,669,583	£1,252,009	
Mean	92,055	43,173	0.45

Source: Select Committee of the House of Lords on the State of the Coal Trade, P.P. (1830), vol. 8, 29; P. Cromar, "Economic Power and Organisation: The Development of the Coal Industry on Tyneside, 1700–1828" (Ph.D. dissertation, Cambridge, 1977), App. 6.1, 173.

coalfield as the debt burdens of collieries rose along with rising capital requirements.[36]

Such very significant capital investments made coal entrepreneurs particularly sensitive to working costs, a point of great significance for the ideology of the trade.[37] Costs were roughly divided between two principal areas: that of actual production and that of delivery, or "leading," to the fitter or other initial purchaser. Production costs were generally composed of wages, underground transport, repair and maintenance, horses, and rents and taxes. Leading costs were composed almost exclusively of wages, repair and maintenance, and rents.

Costs, however, varied not only between individual pits but within the same pit as well. The costing of the eight working pits at the Lambton collieries in 1835 revealed that while coal could be brought to bank at 3s. 8d. per Newcastle chaldron at Littletown Colliery, the same process cost 5s. 1d. at Cocken Colliery. At Newbottle Colliery, the Dorothea Pit raised coal to the bank from the Maudlin Seam at 4s. 11d. per Newcastle chaldron, but it cost 5s. 11d. to raise coal from the deeper Hutton Seam in the same pit.[38] Each pit thus had a unique cost structure based on different wage rates that applied to different working conditions and upon different haulage distances to the surface.[39]

A broad range of other costs, including those for engines, craftsmen, horses, and rents, further contributed to the variations in costs. Table 1.4 breaks down working costs for each of the four Lambton collieries in 1835. The most striking evidence to be derived from these costings is the high proportion of working costs taken up by underground wages. Like most northeastern collieries in the first third of the nineteenth century, the Lambton collieries' underground wage bill on average amounted to over 50 percent of total working costs.[40] Moreover, the enormous variety of expenditures is also indicative of the vast amounts of cash needed to keep the mines operating. In the first six months alone, the Lambton mines expended over £38,000, of which over £20,000 was needed for underground workers' wages.

Other than labor, leading, or transport, charges were the most significant costs incurred by collieries, and it is in the transport of coal that technological

---

[36] Kenwood has estimated capital–output ratios at 0.72 for the 1840s, a sizable increase. See Kenwood, *Capital Formation*, 72. Unfortunately, capital–output ratios are not disaggregated by region in the new standard history by Roy Church, *The History of the British Coal Industry, 1830–1913: Victorian Pre-eminence* (Oxford: Clarendon, 1986).

[37] See Chapter 2.

[38] D.C.R.O., NCB I/JB/2084.

[39] On wage rates, see Chapter 5.

[40] Flinn, *History of the British Coal Industry*, 292–3.

Table 1.4. *Working costs of Lambton collieries, 1835 (percentage of total costs)*

	Colliery (No. of pits)			
	Cocken (1)	Lumley (3)	Littletown (2)	Newbottle (3)
Underground wages[a]	42	53	59	55
Craftsmen wages and materials[b]	4	8	7	6
Engines and wages[c]	3	3	4	3
Leading charges[d]	21	11	12	9
Horses	8	6	9	9
Colliery rents	9	12	—	10
Wayleave and damaged ground rent	10	5	6	5
Taxes and cesses	3	2	1	3
Interest	—	—	2	—

[a]Working and bringing to bank.
[b]Smiths, masons, wrights, stockkeepers, saddlers; timber, iron, lead, candles, oil, nails, tiles, bricks, etc.
[c]Boilers, iron, timber, oil, grease, ropes, leather; enginemen and firemen's wages.
[d]Charges assessed individually for each pit to main railway terminus at Philadelphia; general charges for all coal to Sunderland apportioned equally.
Source: D.C.R.O., NCB I/JB/2084–8.

innovations were most eagerly adopted. Waggonways, railways, engines, and horses all represented substantial capital investments.[41] Fixed stock alone on the rail line between Philadelphia and Sunderland owned by the earl of Durham, for example, was valued at £57, 197 in 1835, and movable stock was estimated to be worth an additional £20,455.[42]

It was common, moreover, for coal owners to transport competitors' coal for relatively reasonable rates or to share railway stock. In 1814, Arthur Mowbray, in an attempt to modernize the Vane-Tempest collieries, presented the executors of the estate with a variety of expansion options including ways to limit capital outlays on a new railway. In one option, the estate could finance the building of a short branch line into an existing railway owned by John Nesham, another local coal owner.[43] Upon further investigation, however, William Vizard, Lady Frances Anne Vane-Tempest's solicitor, noted that J. G. Lambton had also expressed a willingness to share

[41] Ibid., 146–59; M. J. T. Lewis, *Early Wooden Railways* (London: Routledge & Kegan Paul, 1970); Edward Hughes, *North Country Life in the Eighteenth Century: The North-East, 1700–1750* (Oxford: Oxford University Press, 1952/1969).
[42] D.C.R.O., NCB I/JB/2084–8.
[43] D.C.R.O., Londonderry Papers, D/Lo/B 37(2).

the cost of building a longer railway to Sunderland.[44] Throughout the 1830s, the marquess of Londonderry leased space on Lord Durham's railway line between Pittington Colliery, Bourn Moor, and Philadelphia and was charged one pence per Newcastle chaldron. The two lords also shared railway lines to Sunderland. These lines transported coal at a mutually accepted price of 6.09d. per chaldron and moved over 45,000 tons of coal for each concern in the first half of 1835.[45]

The practice of utilizing rail stock to carry the coal of competitors, moreover, seems to have lowered barriers to entry to the trade in certain circumstances. In the 1830s, for example, North Hetton Colliery Company shipped its coal along Lord Durham's line between Pensher and Low Lambton staith on the Wear.[46] Nevertheless, the maintenance of these barriers to entry, or at least the control of these barriers, was perceived by the large coal owners as an essential part of their dominance in the trade. When the provision of public railways after 1830 threatened to lower barriers to entry significantly and thereby wrest an important source of power in the trade away from the landed coal entrepreneurs, the coal lords fought a long and losing battle to protect their rights from what they claimed was state interference with the prerogatives of private property.[47]

Fixed-capital demands, therefore, were exceedingly heavy in the coal industry during the initial phase of industrialization. Barriers to entry could be lowered by the practice of leasing lands both above and below ground as well as by hiring private rail stock for haulage, and, by the 1830s and 1840s, the provision of public railways for coal transport. Scant evidence also exists that barriers to entry may have been lowered by the purchase of used capital equipment. For example, North Hetton Colliery in 1832 examined Lord Durham's hoist machinery for possible purchase but found it could not be adapted to their needs.[48] This may have been a common problem. Yet acquiring, maintaining, and extending fixed-capital stock, however burdensome, was only one aspect of mining. Securing the necessary cash and credit to carry on operations was a further problem for enterprise during the early industrial era.

Cash and credit

The scale of the working costs of a large mining operation necessitated immense quantities of cash. Unlike the Yorkshire textile industry recently

[44] Ibid., D/Lo/B 43(iii).
[45] D.C.R.O., NCB I/JB/2084.
[46] Ibid.
[47] Kenwood, *Capital Formation*, 78.
[48] N.C.L., Matthias Dunn's Diary, 4 December 1832.

studied by Hudson in which liquidity could be limited by the regular extension of credit and the prevalence of the truck system of wage payments, the coal industry demanded not only high fixed capital costs but constant liquidity as well.[49] The aristocratic or gentry status of many coal owners no doubt helped them to tap the local capital market. However, the links forged by the coal trade between the Northeast and London allowed access to wider, indeed national, capital and was of great significance to the long-term growth of the northern coal industry.

The papers of the great Tory coal owner, the third marquess of Londonderry, reveal the degree to which liquidity was required for a large coal concern. The Londonderry's colliery accounts of 1820–5 credit an average of £156,380 annually to cash received and cash due from fitters. Over the same period, the colliery cash books paid out an average of £113,261 per annum, or over 72 percent of cash received.[50] Most cash payments went to wages and materials, of which the former was by far the most significant. Colliery wage bills averaged £9,600 per month in the last five months of 1825, although it seems that monthly colliery pays were expected to total approximately £8,000.[51] Significantly, the truck system was not widely practiced in the northern coalfield, and the pitmen lived in a fully cash-based local economy.[52] This was so much the case that during the banking crisis of 1825, it was noted: "The pitmen don't like Bank of England notes so well as provincial notes from the Banks in which they have confidence."[53] Thus, large amounts of cash preferably from local banks was a constant necessity for firms operating in the Northeast.

Until the second quarter of the nineteenth century, the Tyne and Wear coal fitters, or independent shipping agents, were the crucial link in a circuit of capital that transferred these huge sums to the Northeast through bills of exchange. Fitters contracted the loading of the colliers and, more importantly, arranged the sale of the coal to the shipmasters. Once in London, the shipmaster arranged for the off-loading and sale of coal through a factor on the London Coal Exchange who in turn was responsible for the sale of the coal to a range of retailers.[54] A coal owner's revenue was ultimately

[49] Hudson, *Genesis of Industrial Capital*, 109–54.
[50] D.C.R.O., Londonderry Papers, D/Lo/B 81 (3, 15, 30, 32–5).
[51] D.C.R.O., Londonderry Papers, D/Lo/B 142(148–9), Buddle to Londonderry, 14 August 1825; 23 March 1826.
[52] B. R. Mitchell, *Economic Development of the British Coal Industry, 1800–1914* (Cambridge University Press, 1984), 237.
[53] D.C.R.O., Londonderry Papers, D/Lo/C 142(148), Buddle to Londonderry, 28 December 1825.
[54] Flinn, *History of the British Coal Industry*, 267–79; Raymond Smith, *Sea Coal for London* (London: Longman Group, 1961).

dependent upon the receipt of the fitters' bills of exchange. At any one time, upwards of £20,000 could be due to a large coal owner. At the Londonderry collieries between 1819 and 1825, fitters annually owed an average of £23,070 at year's end; actual receipts from fitters averaged over £10,300 each month during the same period.[55]

The transfer of such substantial amounts required a dense network of personal and institutional security that was generally local in origin and rooted in the professional and mercantile interests of the community. On the Tyne, as Flinn has described it, fitters acted as "independent agents" who were retained by several coal owners and remunerated by standing fees or "fittage" charges.[56] On the Wear, however, such was not the case. Fitters, like coal miners, were contracted under the terms of a "bond" often to only a single coal owner. The bond recorded the security offered to the producer in terms of both money value and a list of local guarantors. Many of the bonds entered into by Lord Stewart (later Lord Londonderry) in the early 1820s varied in value from between £200 and £600. Yet it is apparent that there was also a significant degree of latitude allowed to some fitters. Isabella Lees, for example, one of two women who were fitters on the Wear, was not required to offer any security in 1814.[57] At the other extreme, however, William Hayton, one of the five largest fitters on the Wear, offered security in the amount of £5,000 in 1821.[58]

Even more interesting is the social origin of those who offered their own personal security for the bond. Among the larger fitters, gentlemen, ship-owners, and local colliery viewers were common. Thus, William Hayton's bond was secured by William Stobart, Jr., a local colliery viewer, and by Robert Thompson, a shipowner from Bishopwearmouth. In 1820, Hayton vended over 80,000 tons of coal from the Wear. Smaller fitters, such as Joseph Clark who vended only about 18,000 tons of coal in 1820, offered the security of Henry Moon, shipowner, and John Clark, a butcher. Other small fitters offered the security of local grocers, brickmakers, and sur-geons.[59] In at least one case, we know that the fitters were neither local nor independent small businessmen or women. Catto and Co. was not one of the larger players among Wear fitters; they vended only 22,000 tons of coal in 1820. Yet they were a partnership of eleven members operating out of Aberdeen who transported coal north to Scotland rather than to London, the principal and most profitable route. Nonetheless, this is an important

[55] D.C.R.O., Londonderry Papers, D/Lo/C 81.
[56] Flinn, *History of the British Coal Industry*, 269.
[57] D.C.R.O., Londonderry Papers, D/Lo/B 125.
[58] Ibid., D/Lo/B 125(17).
[59] Ibid., D/Lo/B 128(8).

indication of the extent to which the northern coal trade was integrated into a wider market and the extent to which national capital was increasingly impinging upon the northern coal trade.

At times, fitters could play a very significant role in the marketing of coal. During periods in which the coal cartel, the so-called Limitation of the Vend,[60] was effectively distributing quota allotments to individual collieries, fitters seem to have had little trouble in regularly engaging shipmasters to purchase and transport the coal. However, during periods of rapidly declining prices or open competition, fitters found that shipmasters were reluctant to purchase coal and risk a loss when the coals came to market. During these times, an aggressive body of fitters who could secure shipping might make the difference between a firm's survival and its failure. Thus, in 1830, Buddle explained to Lord Londonderry that "the Market is completely over-done with Walls-end [coal] – the Ship-owners *positively refuse* to take them, so that we are entirely thrown upon the strength of our Fitters – the same as in an open Trade."[61] Similarly, during the banking crisis of 1825, falling prices on the coal market in London induced the shipowners to lay up their ships in port "as fast as they arrive." Here again, Buddle noted that the Londonderry collieries "have therefore the strength of our Fitters only to rely on for forcing the Vend."[62]

In the early nineteenth century, especially as coal owners increasingly built staithes below the Tyne Bridge and at Sunderland that could deposit coal directly into the ship's hold via "spouts,"[63] many coal owners dispensed with fitters and engaged shipmasters directly. When J. G. Lambton's (later Lord Durham's) Newbottle spouts came into operation in 1825, for instance, he dismissed his fitters.[64] As early as 1820, a list of Wear fitters and other purchasers showed that two of the nine Wear coal owners, W. M. Lamb & Co. and John Nesham & Partners, fitted all or nearly all of their own coals.[65]

Fitters, however, also could serve other financial functions. They were, for example, sometimes used as sources of short-term loans, and their bills were employed as security for cash advances. In 1832, Buddle secured a loan of £500 from some of Lord Londonderry's fitters who "agreed to forego the receipt of £500 themselves for *one* month."[66] Both in 1826 and in 1831,

---

[60] See Chapter 2.

[61] D.C.R.O., Londonderry Papers, D/Lo/C 142(514), Buddle to Londonderry, 11 June 1830.

[62] Ibid., 142(148), Buddle to Londonderry, 22 December 1825.

[63] See Chapter 2.

[64] Flinn noted the dispute between Lambton and his fitters but was unsure as to the result. However, Buddle notes that Lambton began fitting his own coals in October 1825. See D.C.R.O., Londonderry Papers, D/Lo/C 142(148), Buddle to Londonderry, 15 October 1825.

[65] Ibid., D/Lo/B 307(3–4).

[66] Ibid., D/Lo/C 142(855), Buddle to Londonderry, 27 June 1832.

Londonderry's banker, Edward Backhouse, offered to advance the collieries several thousand pounds but only if the fitters' bills of exchange were placed in his hands directly.[67] In 1831, Londonderry refused, and Backhouse reacted "as if I had stopped him on the High-way and demanded his purse."[68] But considering Londonderry's financial difficulties during that strike year, he ultimately was forced to accede to the banker's demands.[69]

It also was not uncommon for fitters to be the source of the backward migration of capital from fitting into the direct exploitation of coal mines. In the Hetton Coal Company, two of the eleven initial partners, William Hayton and Thomas Horn, were fitters from Bishopwearmouth.[70] Both Hayton and Horn held two shares in the company valued at £250 a share. Similarly, in 1831, William Bell, a Wear fitter, took over South Moor Colliery from Lord Ravensworth and, by securing a wayleave from Lord Durham, shifted that colliery from shipping on the Tyne to the Wear.[71] Thus, while it is true that, as Ward contends, skilled mining engineers were prominent in the expansion of the coal industry in the nineteenth century, the same could be and should be said of fitters.[72]

The Londonderry collieries were unique to a certain extent in that they embarked upon the construction of Seaham Harbor, an enterprise that swallowed enormous sums of capital. However, their constant recourse to country banks to discount fitters' bills of exchange and to secure short-term advances in order to meet wage bills probably was not unusual. Country banks played a significant role in the financing of Northeast coal operations.[73] As often was the case, the short-term loans offered by country banks were often converted into long-term debts. Constant overdrafts and extensions of credit allowed Lord Londonderry to accumulate a debt of over £20,000 to the country bank Ridley & Co. of Newcastle in the mid-1820s.[74] Nor was Ridley's of Newcastle the only country bank to feel Lord Londonderry's pinch. As noted above, throughout the 1820s and 1830s, he was constantly at odds with the Quaker bankers Backhouse & Co. of Sunderland and Darlington.

[67] Ibid., D/Lo/C 142(149), Buddle to Londonderry, 2 January 1826.

[68] Ibid., D/Lo/C 142(734), Buddle to Londonderry, 15 June 1831.

[69] Ibid., D/Lo/C 142(739), Buddle to Londonderry, 23 June 1831.

[70] Ibid., D/Lo/B 309(19). The list of original partners is reproduced in Sill, "Landownership and Industry," although the fitters are not identified as such.

[71] D.C.R.O., Londonderry Papers, D/Lo/C 142(738), Buddle to Londonderry, 21 June 1831.

[72] J. T. Ward, "Landowners and Mining," in Ward and Wilson, *Land and Industry*, 72.

[73] On country banks, see L. S. Pressnell, *Country Banking in the Industrial Revolution* (Oxford: Oxford University Press, 1956); François Crouzet, "Introduction" and "Capital Formation in Great Britain during the Industrial Revolution," in François Crouzet (ed.), *Capital Formation in the Industrial Revolution* (London: Methuen & Co., 1972); and Rondo Cameron et al., *Banking in the Early Stages of Industrialization: A Study in Comparative Economic History* (New York: Oxford University Press, 1967).

[74] D.C.R.O., Londonderry Papers, D/Lo/C 142(147), Buddle to Londonderry, 13 June 1825.

In 1829, he maintained outstanding loans of at least £10,000 with them.[75] Perhaps in desperation, Backhouse & Co. offered to advance a further £10,000 to the Londonderry collieries in 1831 in return for complete control and working privileges to Londonderry's Grange Colliery.[76] This bank, for one, was quite willing to involve itself directly in the management of an active enterprise.[77]

Many aspects of the variegated capital markets that have been reconstructed for Lancashire and Yorkshire during the early stages of industrialization apply equally to the northern coal industry.[78] Mortgages in both the local and London market seem to have been a frequent source of long-term capital. In 1829, Lord Londonderry had a £35,000 mortgage on his Seaham estate.[79] Two years later, in June 1831, Backhouse & Co. threatened to take possession of Londonderry's freehold Pensher Colliery under the terms of their mortgage deed.[80] In the early 1830s, several mines in the region lapsed into the hands of bankers, most notably Silksworth and Ludsworth Colliery, which was taken over by Gurney's, a Norwich country bank.[81] This is a striking indication of the extent of country bank investment outside of the purely local or London market, but Gurney's connection to the Northeast was a long-standing one. As early as 1776, Bell & Co. of Newcastle, a substantial country bank, had an account with Gurney's to settle transfers of notes and drafts.[82]

In the case of the Londonderrys, most long-term loans and mortgages were secured in London.[83] Among the marquess's London bank creditors were Coutts & Co. and Drummonds, both of which held mortgages on portions of the estate worth nearly £25,000 in 1829.[84] Smaller London banks, such as Rundell & Bigge, also held mortgages on the main Wynyard estate in Durham.[85] Loans secured through personal contacts and family relation-

---

[75] Ibid., D/Lo/B 310(7).

[76] Ibid., D/Lo/C 142(724), Buddle to Londonderry, 8 June 1831; D/Lo/C 142(734), 15 June 1831; D/Lo/C 142(746), 2 July 1831; D/Lo/C 142(749), 6 July 1831.

[77] Cameron, *Banking*, 56–8, stresses the relatively widespread participation of bankers in industry; while, conversely, Pressnell, *Country Banking*, 15–28, emphasizes the industrial origins of many country bankers.

[78] Hudson, *Genesis of Industrial Capital*, chaps. 5–7; B. L. Anderson, "The Attorney and the Early Capital Market in Lancashire," in J. R. Harris (ed.), *Liverpool and Merseyside: Essays in the Economic and Social History of the Port and Its Hinterland* (New York: Kelley, 1969).

[79] D.C.R.O., Londonderry Papers, D/Lo/B 310(7).

[80] Ibid., D/Lo/C 142(729), Buddle to Londonderry, 11 June 1831.

[81] Ibid., D/Lo/C 142(719), Buddle to Londonderry, 4 June 1831.

[82] Pressnell, *Country Banking*, 130.

[83] A similar case for the economy in general is made by Crouzet, "Introduction."

[84] D.C.R.O., Londonderry Papers, D/Lo/C 142(782), Buddle to Londonderry, 6 February 1832; D/Lo/B 310(7); Sturgess, *Aristocrat in Business*, 83.

[85] D.C.R.O., Londonderry Papers, D/Lo/C 142(779), Buddle to Londonderry, 13 January 1832; D/Lo/C 142(780), 3 February 1832.

ships were also significant in the financing of the northern coal industry. In 1825, Lord Londonderry's largest single debt was owed to Sir Edward Banks, who advanced him £30,000 on the mortgage of Holdernesse House, Londonderry's Park Lane residence.[86] By 1829, Londonderry still had £28,000 outstanding to Sir Edward in addition to £8,000 borrowed from his brother-in-law, Sir Henry (later First Viscount) Hardinge, and a further £5,000 owed to "Dr. Frank."[87]

Finally, the members of the landed class who engaged in the northern coal industry were unique for the quite obvious reason that their income was always supplemented by proceeds from their estates, thus closing the circle of land and industry. In the case of Lord Londonderry, the vast amount of debts he was accumulating in the 1820s and 1830s was mitigated partially by the fact that he continued to receive an annual income of approximately £22,000 from his English as well as his Irish estates.[88] Thus, in Buddle's plan to liquidate Londonderry's debts drawn up in 1825, "the proceeds of the Irish property [were] to be appropriated to Ld. L-d's private purposes and for the maintenance of the Family, i.e. Lord L. to live on the proceeds of the Irish property."[89] At approximately £12,000 per annum, one could continue to live well enough.

In conclusion, therefore, the origin, forms, and sources of capital for the development of modern industry in the Northeast were varied. Capital for the mining industry came from both local and national sources, while the principal methods of tapping these markets included personal loans, temporary bank advances, short-term overdrafts converted into long-term loans, and mortgages. The social position of a noble entrepreneur like Lord Londonderry undoubtedly enabled him to exploit more fully the several reservoirs of capital. Yet the scale of capital requirements, in both its fixed and circulating forms, created an economic context that provided the motivation to more fully control labor and commodity markets as well as a series of obstacles to the enforcement of that control. The "web of credit" in fact ensnared capital not only when it balanced its account books but when it confronted the marketplace as well.

Markets and output

Once coal was delivered to the staithes, or quays, it generally passed out of the hands of the coal owners and into the hands of a series of middlemen. There is no doubt that throughout the first half of the nineteenth century,

---

[86] Sturgess, *Aristocrat in Business*, 49.
[87] D.C.R.O., Londonderry Papers, D/Lo/B 310(7).
[88] Ibid., D/Lo/B 310(7); D/Lo/C 142(510).
[89] Ibid., D/Lo/B 310(4).

London was considered the principal market for Northeast coal. Even into the railway age, the comparative cost advantage provided by the river and coastal transport of coal secured the dominance of northern coal in the London market. Significantly, the cartel restrictions of the Limitation of the Vend applied only to coal sold in the London market. Other markets for coal were unrestricted and competitive. However, as early as the second quarter of the nineteenth century, there began a gradual reorientation of the northern coal industry away from London household consumption and toward overseas exports and local industrial consumption.[90] Between 1800 and 1870, London's consumption of total Northeast output declined from approximately 30 percent to just over 10 percent even while the capital's consumption of Northeast coastwise shipments fell from only about 55 percent in 1800 to approximately 52 percent seventy years later.[91] It is thus quite apparent that, after midcentury, at the same time that London continued to be the largest coastal market for Northeast coal, local entrepreneurs were increasingly diversifying their markets. This was done first by expanding their shipments of household coal along the eastern coast, then by expanding foreign exports, and, finally, by meeting the growing demands of the local iron and metal working industries.

One indication of the new interest among northern coal owners in the consumer market outside of London was the competition between Lord Londonderry and Lord Durham for control of the East Anglian market. In 1853, Lord Londonderry was losing his market share in Northampton, Cambridge, Bedford, and Norfolk both to Lord Durham, who had adopted a screw steamer to transport coal, and to Yorkshire and Derby coals, which were being delivered into the region after the junction of the Great Northern Railway with the Norfolk and East Anglian railways.[92] At Peterborough, the best coals from Derby were selling at 11s. 6d. per ton and those from Yorkshire at 12s. 6d. a ton. However, Londonderry's best coals could not be delivered from Seaham for less than 16s. per ton. A similar story was told for Cambridge, where inland coals were selling at 16s. a ton, while Londonderry's Eden Main coals were selling at 19s. 4d. if transported through Wisbech or 20s. 4.5d. if through King's Lynn.

The coastal market for coal was clearly demarcated into price territories based upon access to low-cost transportation. Where inland coals transported by rail penetrated the market, they almost invariably drove out coal transported by sea from the Northeast. This was the case in the area east of

[90] Mitchell, *Economic Development*, 15–19.
[91] Flinn, *History of the British Coal Industry*, 217, 273–4.
[92] For this and the following, see D.C.R.O., Londonderry Papers, D/Lo/B 73, "Report on the Coal Trade of the Eastern Counties, October 1853."

Cambridge: "From Cambridge inland Coals are carried eastward to Newmarket by Railway and until the Railroad from Newmarket to Bury St.
Edmunds is opened, they go no further, for then they will come into competition at Bury with Seaborne Coals from Ipswich, which now have a
monopoly and the price is 24s/ a Ton." However, where local navigable
waterways existed, the market territory of northeastern coal was extended
at the expense of their inland competitors. Thus, it was reported, while there
was little hope of expanding the market east of Cambridge, there was considerable room for growth in the area to the west and south of Peterborough
where "we have a navigable river all the way to Northampton, and South
of St. Ives and Huntington to Bedford."

Unlike the London market where the trade in coal was highly specialized,
marketing in the coastal markets was left to an almost infinite variety of
merchants with only a part-time interest in the coal trade. It was common
at King's Lynn, for example, for shipowners also to arrange the marketing
and distribution of coal, but this was carried on as well by William Clifton
and Son, wine and spirit merchants, and Mr. Bowker, a corn factor. At
Wisbech, shipowners were once again prominent marketing agents as was
G. Presh of Sutton Bridge, a timber merchant. In this port, it was apparently
common for shipowners to be highly diversified on a very small scale. Thus,
John Collins of Wisbech was a shipbuilder, shipowner, and operated a large
glass and earthenware shop; J. M. Patrick owned a small coasting vessel and
was also a watchmaker.

Both Lord Londonderry and Lord Durham aggressively pursued these
new markets even though they constituted a relatively small proportion of
coastwise shipments. By the early 1850s, Lord Durham's screw steamer
could make a voyage a week between Sunderland and Wisbech carrying six
hundred tons of coal compared with an average collier's load that has been
estimated at around three hundred tons or less.[93] This allowed Durham's
best coals to be sold at from threepence to one shilling below that of Lord
Londonderry's. Londonderry's response was typical of his rash and aggressive personality as well as the decided competition for these markets: "We
should instantly order a Screw Steamer," he wrote in the margin of this
report. "Mr. Hindhaugh is hereby remanded to do this and to let Lord L
know the necessary payments to the Builder. To be called the Londonderry
Screw Steamer."[94]

The continuous and largely successful pursuit of new markets after 1830

---

[93] William J. Hausman, "The English Coastal Coal Trade, 1691–1910: How Rapid Was
Productivity Growth?" *Economic History Review*, 2nd ser., 40, no. 4 (1987), 593.
[94] D.C.R.O., Londonderry Papers, "Report on the Coal Trade of the Eastern Counties, October 1853," D/Lo/B 73.

is evidenced by the growth of total coal output from the Northeast. Flinn's revised estimates of coal production in the eighteenth and nineteenth centuries show that total coal output from the Northeast rose from nearly 3 million tons in 1775 to just under 7 million tons by 1830.[95] According to Church, total coal output per annum from the Northeast then rose to 15.2 million tons by 1850–5 and to over 54 million tons in 1910–13[96] (see Figure A.1 of the Appendix).

However, it is important to emphasize once again that although the London market was probably the most profitable and perceived by coal owners to be the most secure, the capital accepted an increasingly smaller share of total northeastern output through the nineteenth century. Thus, the economic rationale of the coal cartel, holding up prices in the London market, continued to play a smaller role in the financial prospects of many northern collieries. This does not mean that the cartel collapsed in 1844 because it had simply outlived its usefulness when the coal industry entered the modern industrial world. On the contrary, it still performed essential functions as a lobbying group and bargaining unit during miners' strikes. However, the greatest coal owners, the Londonderrys and Durhams, for example, became ever more aware of the fact that their profits could be sustained by shifting a greater proportion of their sales into the unregulated coastal and foreign markets and to the developing iron industries. Hence, overseas exports rose as a proportion of total output from approximately 4 percent to nearly 13 percent between 1816 and 1840 and the consumption of coal by railways, steamships, and iron works from nil to approximately 7 percent of output in 1840[97] (see Figures A.2 and A.3 of the Appendix).

One need not fully accept, therefore, Paul Sweezy's argument that there was an inherent logic in the organization of coal production that promoted the expansion of capacity while at the same time undermining the cartel's cohesiveness.[98] Solely in terms of the London market, it may be true to say that industrial efforts to restrict production were unstable essentially because supply outstripped demand, and in an industry saddled with extremely high proportions of fixed costs, competitive pressures might prove to be inexorable. Yet London was increasingly less important to the market position of the leading firms and absorbed smaller portions of their output. Moreover, before the midcentury collapse of the cartel, as we shall see, coal owners

---

[95] Flinn, *History of the British Coal Industry*, 26–7; Sidney Pollard, "A New Estimate of British Coal Production, 1750–1850," *Economic History Review*, 2nd ser., 33, no. 2 (1980), 223.

[96] Church, *History of the British Coal Industry*, 3. Mitchell's statistics are slightly different; see Mitchell, *Economic Development*, 7.

[97] Mitchell, *Economic Development*, 16–17.

[98] Paul M., Sweezy, *Monopoly and Competition in the English Coal Trade, 1550–1850* (Cambridge, Mass.: Harvard University Press, 1938), 115–6.

were not always economically rational. Generally, they were more determined to secure their profits through fixing prices rather than through lowering costs and expanding markets. This, in part, was a legacy of the corporate mentality of the trade as well as its instrumental definition of the laws of the market.

2

The perception of the market and industrial policy

Laws in economics are determined by their opposite, lawlessness. The true law of economics is *chance*, and ... people arbitrarily seize on a few moments and establish them as laws.

Karl Marx, *Excerpts from James Mill's* Elements of Political Economy (1844)

In his brief classic *The Industrial Revolution*, T. S. Ashton rightly warned historians and students alike that while the era of industrialization may properly be characterized as a period of laissez-faire, it cannot with equal justice be considered a time of rampant individualism.[1] Both capital and labor were imbued with a "corporate sense" that was far more pervasive than the atomistic caricature portrayed by political economists.[2] Without doubt, this corporate sense extended to the northern coal owners, who viewed the market far less as an invisible hand allocating supply and demand and setting prices than as a potent threat to profits and a tangible bludgeon with which to punish erstwhile competitors. Both industry-wide and at the level of the individual firm, the organization of coal production revolved around efforts to raise the market price of coal, thus subverting the market, and not to encourage effective competition.

The efforts to raise the market price of coal resulted in the formation of the cartel of energy producers, widely known as the Limitation of the Vend. The cartel operated to allocate quotas exclusively in the London coal market. Despite the apparent identity of interests among coal owners, the cartel was seriously weakened by conflicts generated by differences of scale, especially

[1] T. S. Ashton, *The Industrial Revolution, 1760–1830* (Oxford: Oxford University Press, 1948; repr. 1980), 88–97.
[2] The prevalence of combination and cooperation among early industrialists was similarly pointed out by J. H. Clapham in *An Economic History of Modern Britain: The Early Railway Age, 1820–1850* (Cambridge University Press, 1959), 198–205. On the history of the corporate idiom in France, see William H. Sewell, *Work and Revolution in France: The Language of Labor from the Old Regime to 1848* (Cambridge University Press, 1980).

those that arose among the largest members in the trade.[3] At this level, the coal owners' support of the cartel was both essential and ephemeral. The largest coal owners accepted the production limits imposed upon them by the cartel only as long as they did not prove to be either an impediment to their profits or a threat to their control of the market. Once these limits were breached, however, the great coal owners trained the fire of the free market upon their opponents in the hope of driving them from the field. Therefore, both the efforts of cartellization and the recurring bouts of free trade in the northern coal industry during the early nineteenth century were signals of the attempts of the largest producers to control the market.

Consequently, political economy and the dictates of the market were not strictly amenable to the organization of production in the coal trade. In fact, in the public presentation of their trade, coal owners struggled to manipulate the discourse of political economy, especially those concepts embodied in the dichotomies of supply and demand, monopoly and competition, and regulation and free trade. Ultimately, however, their public prescriptions were less important than their political power and influence. Parliament never did act to break up the cartel, even if this had been possible, and the Vend disintegrated in the mid-1840s under pressures from both within and outside of the regional coal industry.[4]

Supply and demand

Despite the dominance that political economy had established over the terms of economic and social discourse by the 1830s, the northern coal owners assiduously tried to undermine its application to the coal industry.[5] In particular, they argued that the coal industry was so circumstanced that both classical supply-and-demand and monopoly theories were inadequate as well as inaccurate models by which to judge the effect of industrial organization on consumers. In the end, the trade's language was drawn as much from the vocabulary of custom and privilege as it was from that of political econ-

---

[3] Portions of this chapter have appeared in a somewhat different context in my "Competition and the Size of Firms in the North-east Coal Trade, 1800–1850," *Northern History*, 25 (1989), 235–55.

[4] Church, *History of the British Coal Industry*, 66–9.

[5] On the discourse of political economy, see Gray, "Languages of Factory Reform," 143–79; Joan W. Scott, " 'L'ouvrière! Mot impie, sordide...': Women Workers in the Discourse of French Political Economy, 1840–1860," in Patrick Joyce (ed.), *The Historical Meanings of Work* (Cambridge University Press, 1987), 119–42; and John Smail, "New Languages for Labour and Capital: The Transformation of Discourse in the Early Years of the Industrial Revolution," *Social History*, 12, no. 1 (1987), 49–71. The latter article has been severely criticized by Adrian Randall, "New Languages or Old?: Labour, Capital and Discourse in the Industrial Revolution," *Social History*, 15, no. 2 (1990), 195–216.

omy; as tradesmen, they sought a "fair price" for their product and a "fair profit" for their concerns, both of which could not be secured in an open marketplace or under competitive conditions. Therefore, despite the obvious parallels that exist between the organization of the coal industry during the early stages of industrialization and both monopoly capitalism and modern large-scale enterprise,[6] the roots of the coal industry continued to lay in the commodity relationships and ideology of the era of manufactures.[7]

Since the early eighteenth century, firms in the northern coalfield had acted to restrict production destined for the London market according to the proportional distribution of an estimated total volume of sales, the so-called Vend.[8] The distribution of the Vend was established for each colliery at committee meetings, at first by the proprietors themselves, later, and more often, by their representatives. As coal mining spread south and east through Durham in the eighteenth and nineteenth centuries, separate committees were established for each major river (Tyne, Wear, and Tees) as well as a general coordinating committee. The establishment of just proportions was never an easy task since it depended on judging the quality and market price of the coal being worked in conjunction with the colliery's productive capacity and operating costs. Coal owners invariably sought the largest allotment possible, and the setting of each colliery's quota usually precipitated disputes.

Various methods of enforcement and enticement were developed in the attempt to maintain industrial self-discipline. In 1711, for example, participants undertook the joint management of the failing Stella Colliery and assumed the burden of its debts to keep the concern in the cartel.[9] Such ad hoc measures, however, were made more regular after 1829, when sophisticated administrative machinery was developed whereby independent inspectors and referees recommended allotments and a grievance procedure was established. Nonetheless, while very small concerns could operate outside of the cartel (in 1836, there were twelve unregulated collieries on the three rivers),[10] the support of the major concerns was imperative. This was

---

[6] See Sweezy, *Monopoly and Competition*, and Cromar, "Economic Power and Organization: The Development of The Coal Industry on Tyneside, 1700–1828" (Ph.D. dissertation, Cambridge University, 1977), or Flinn, *History of the British Coal Industry*, and Sidney Pollard, *The Genesis of Modern Management* (Cambridge, Mass.: Harvard University Press, 1965) for contrasting evaluations of the "advanced" nature of the British coal industry during this period.

[7] Berg, *Age of Manufactures*, 69–91, provides a critical assessment of theories of manufacture.

[8] Sweezy, *Monopoly and Competition*, 37–45; T. S. Ashton and J. Sykes, *The Coal Industry of the Eighteenth Century* (Manchester: Manchester University Press, 1929), 211–25; Hughes, *North Country Life*, 151–257; Cromar, "Economic Power and Organization"; and Flinn, *History of the British Coal Industry*, 254–67.

[9] Hughes, *North Country Life*, 191.

[10] *Report from the Select Committee Appointed to Inquire into the state of Coal Trade, as respects the*

because the productive capacity of the largest concerns could threaten the stability of prices at market whereas that of the smaller collieries could not. In 1828, the cartel collapsed when Lord Durham withdrew his support; Lord Londonderry's withdrawals in 1832 and 1844 had the same effect. The coal owners' cartel, therefore, was dominated by the largest mining interests who led pricing decisions and established industrial policy.

The cartel came under attack from city interests and free traders several times during the first half of the nineteenth century but particularly so in 1836. At that time, Joseph Hume chaired a select committee appointed in response to a petition from London residents who claimed that the cartel acted to sustain prices and prevent competition "in direct opposition to the principles of free trade and open competition adopted by the Legislature."[11] Hume, naturally, was determined to prove that the cartel constituted a monopoly acting against the best interest of consumers. The Select Committee, however, was stacked with representatives of the trade's interests – Hedworth Lambton, Joseph Pease, William Bell were coal owners and members of the committee – and they defended the organization of the coal trade by subtly undermining political economy's claims to both logical explication and economic efficiency.

Free trade, it was argued by the representatives of the northern coal industry, was inimicable to the consumer's best interest. Open competition ultimately drove high-cost producers out of production and consolidated the industry into the hands of a few wealthy individuals. R. W. Brandling, the chairman of the coal owners' cartel, testified that free trade would lower market prices only briefly, after which the few remaining firms would be in the position to raise prices once again. Free trade, he said, would operate beneficially for the consumer only "for a certain time":

> The collieries that were the least able to bear it [competition] would be driven out of the trade; many of the collieries would be laid in, and some would not be brought into the market again. After that had continued for a certain length of time, I conceive that the collieries that survived would naturally raise the price of their coals.[12]

John Buddle, the nationally respected mining engineer, similarly argued against the ultimate efficacy of unrestrained competition. Buddle supported the paradox that competition engendered monopoly. Moreover, Buddle, like

Supply of Coal into the Port of London, and the Adjacent Counties, from the Rivers Tyne, Wear, and Tees, and Other Places (hereafter, *Select Committee on the State of the Coal Trade into the Port of London*), P.P. (1836), vol. 11, 78.
[11] *Select Committee on the State of the Coal Trade into the Port of London*, P.P. (1836), vol. 11, 3; App. no. 1, 207.
[12] Ibid., 11, 23.

other defenders of the northern coal trade, overtly questioned the legislature's role in the determination of market principles.[13] In a revealing exchange, Hume, the radical M.P., sought to emphasize to Buddle that it was Parliament's prerogative to "inquire into the prices and charges on every article that is come to market." However, Buddle resisted that notion, and he was supported by Joseph Pease, who labeled Hume's attitude one of "Parliamentary coercion." Instead, Buddle argued that Parliament's duty was to uphold and ratify traditional modes of production rather than to impose new ones. This argument naturally led him to adopt the language of customary rights and privileges, a corporate vocabulary, rather than the lexicon of political economy. Thus, Buddle said, the current mode of production and distribution was based upon "legitimate" principles;[14] legislative attempts to promote free trade amounted to giving "preference" to some producers, and thus others were "not fairly treated;"[15] and, Parliament had an obligation to "protect" all producers equally.[16]

When further pressed by Hume to explain why the invisible hand should not be placed over the northern coal district, the industrial interests argued repeatedly that supply and demand acted efficaciously only when there was a balance between the two forces. The coal trade was unique among all trades, however, in that supply far exceeded effective demand; consequently, open competition forced prices below a level at which profits were remunerative. (In fact, the disposition of the trade's commodity market was not unique. Trade unions, as we will see, applied the same logic to the organization of the labor market in a similar attempt to capture the definition of the law of supply and demand.)[17] Thus, Hedworth Lambton cleverly got James Bentley, a London coal factor, to admit during the hearings that the coal trade's "greatest peculiarity [was] its excess power of supply."[18] Similarly, in 1830, Brandling argued before an earlier Select Committee that the removal of coastal coal duties would create the demand sufficient to obviate the necessity of a cartel. Six years later, however, when Hume took Brandling over the same ground once again, the cartel's chairman awkwardly testified that the trade's production limits still were needed. The opening of new collieries along the Tees, he stated, had further increased the supply of coal to such an extent that it far exceeded demand: "My expectations at that time [in 1830] have not been realized," Brandling explained. "The demand has

---

[13] Ibid., 123–4.
[14] Ibid., Q. 1801.
[15] Ibid., Q. 1790; Q 1792.
[16] Ibid., Q. 1785; Q 1790.
[17] See Chapter 7.
[18] *Select Committee on the State of the Coal Trade into the Port of London*, P.P. (1836), vol. 11, 75.

not increased in the same proportion with the supply; the powers of supply from the different collieries have increased in a greater proportion than the demand, which had not increased to the extent I expected it would have done."[19]

The coal trade's attempt to undermine the legitimacy of political economy is further evident in the defense of its cartel against the claims of monopoly. Hume and Buddle again locked horns on this issue. Hume was determined to get Buddle to admit that the cartel's regulation of the supply of coal sustained artificially high prices at market.[20] This, therefore, constituted monopoly. However, Buddle cleverly turned classical political economy's monopoly theory against Hume. He suggested that it was impossible to consider the cartel a monopoly because it did not charge monopoly prices. Hume unwittingly admitted that the cartel could raise its prices three or four shillings a ton, but that would attract Scottish and Welsh coal into the London market. That was proof, Buddle exclaimed, that the cartel was no monopoly. In an interesting exercise of antiradical gymnastics, Buddle further reasoned that the cartel's moderate price policy actually encouraged competition. By securing reasonable profits, the cartel had adopted a policy that served to attract entrepreneurs, thus undermining the control of the trade's largest players. Buddle claimed that the cartel set prices that

> secur[ed] to the trade something like a steady renumeration, although not an exorbitant profit, [and] it has been the means of inducing great numbers of individual adventurers to enter into the trade (men of moderate and small capital); and by that sort of competition among collieries, I think it has been the means of preventing either opulent individuals, or large companies, from securing a monopoly of the trade.[21]

The coal trade's paradox thus was complete: Competition created monopoly and monopoly created competition.

Brandling, it should be noted, followed a somewhat different tack in arguing the trade's refusal to be branded as monopolistic. He said, much more simply, that political economy's definition of a monopoly denoted the absence of competition. However, the British coal industry was regionally competitive; Scottish, Welsh, and Yorkshire coals all competed in the market with northeastern coal. Their products were simply priced out of the London market, he implied. Thus, Brandling concluded, "It appears to be a contradiction in terms to say that a monopoly exists where there are two persons who have the same article which they bring to the same market, and between

[19] Ibid., 28.
[20] Ibid., 122.
[21] Ibid., Q. 1755.

whom there is no understanding. In my opinion, it is totally impossible a monopoly should exist under such circumstances."[22]

The final argument by which the representatives of the coal trade rejected the legitimacy of political economy was that of price and price determination. Once more, the industry tended to adopt customary, corporate, and non-market language. According to Buddle, prices were determined by what would afford "something like a fair remunerating price, to cover the expenses of working their coals, and to indemnify them for the investment of their capital and the risk."[23] Buddle's definition, couched in the language of "fairness," interestingly combined a customary definition of profit with a pragmatic element of costs. However, this was not so in the case of Brandling. The chairman of the cartel robbed the language of classical political economy of its most fundamental meaning. Whereas Smith had defined "natural price" as the sum of the elemental costs of land, labor, and capital,[24] Brandling simply defined it as the point at which competition for commodities abated. He explained that the "natural price" of coal was "a price a little below the price that the public can get the same article for elsewhere."[25] "Natural price," therefore, was not "the central price, to which the prices of all commodities are continually gravitating," as Smith defined it.[26] Instead, it was the price at which the laws of the market became impotent.

Perhaps not surprisingly, the only significant area in which the precepts of political economy were accepted with little hesitation was that of industrial relations, yet even here, as will be shown later, the discourse and practice of paternalism was never totally eschewed.[27] Henry Morton, Lord Durham's agent, repeatedly invoked the principles of supply and demand for labor to justify his attempts to reduce wages. In April 1831, for example, he met a group of pitmen from Lambton Colliery on their way to a union meeting and dutifully explained to them that wages must fall because "labour everywhere was in greater abundance than the demand for it, in short that there was a redundant population."[28] He was surprised that the union leaders did not accept this Malthusian logic. Similarly, "Castor," a particularly hostile opponent of Tommy Hepburn's union, argued that union attempts to raise

[22] Ibid., 12.
[23] *Report from the Select Committee Appointed to Inquire into the State of the Coal Trade in the Port of London* (hereafter, *Select Committee on the State of the Coal Trade*), P.P. (1830), vol. 8, 55.
[24] Mark Blaug, *Economic Theory in Retrospect*, 3rd ed. (Cambridge University Press, 1978), 40.
[25] *Select Committee on the State of the Coal Trade into the Port of London*, P.P. (1836), vol. 11, 23.
[26] Smith, *Wealth of Nations*, 65.
[27] See Chapter 4; Robert Moore, *Pit-men, Preachers and Politics: The Affects of Methodism in a Durham Mining Community*, (Cambridge University Press, 1974), 78–92.
[28] Lambton Mss., Morton to Durham, 6 April 1831.

wages violated natural laws and threatened to bring about the apocalypse of the coal trade:

> [Union leaders] can never alter the immutable law of all trade and commerce, that the price of labour, as of every other commodity, must always be in proportion to the scarcity or the abundance of the supply; and if, as they have done, they continue by combination to enforce any artificial price of labor above this natural standard, they infallibly accelerate the ruin of their occupation.[29]

This Malthusian relationship between population and wages, however, was not the only one advanced by industrialists and their supporters during these years. Lord Durham, in fact, anticipated J. W. F. Rowe's findings of the 1920s that the labor market was a very imperfect guide to wages in the coal industry and that "the peculiar economic structure of the industry... necessitate[d] a close correlation between wages and prices."[30] Lord Durham explained to the first anniversary meeting of the Lambton Collieries Benefit Societies in January 1833 that high fixed costs required that wages parallel the movement of prices:

> The rate of wages depends on the price which is given by the public for the article worked. Now, the price of coals is very low, so much so, that little or no profit is made by the coalowner. In many instances he actually loses, and pays the wages of his men out of his capital, not out of his profits. Be assured that if prices rise, wages rise as a matter of course; but that if prices fall wages must also fall.... In no trade is there less regular profit and more steady and permanent expense.[31]

Nevertheless, the influence of political economy is clear: The relationship between wages and prices was immutable, and any attempt to alter these proportions, as Durham later suggested on the same occasion, must lead to further wage reductions and unemployment.

Costs and prices

The failure of the supply and demand of both wages and commodities to regulate coal production, as Lord Durham noted, was fundamentally attributed to the cost structure of the industry. While the industry produced several different grades of coal, it was commonly acknowledged before 1840

---

[29] N.E.I.M.M.E., Bell Collection, vol. 11, 390–1.

[30] J. W. F. Rowe, *Wages in the Coal Industry* (London: P. S. King & Son, 1923), 120–1. Mitchell makes a similar claim that since most costs were "relatively inflexible in the short run... wages had to move more or less *pari passu* with prices." Mitchell, *Economic Development*, 207.

[31] J. Holland, *The History and Description of Fossil Fuel, the Collieries, and the Coal Trade of Great Britain*, 2nd ed. (1841), 301–2.

that profits arose almost solely from the sale of the high-priced household, or "round," coal. John Buddle explained that "in very few collieries [will] the second-rate coal . . . pay for the working . . . the profit entirely depends upon the proportion of high priced coal which each colliery is capable of yielding."[32] Records of the Londonderry collieries bear this out: In 1838–40, the average production costs per ton (exclusive of transport charges that were likely to amount to two shillings per ton) were approximately 7s. 7d.; Stewart's Wallsend, Londonderry's highest priced coal, was sold to the factor at 11s. 6d. per ton. Londonderry's middling grade sold at just under 10s. and small coal was put on board ship at 5s. 6.5d. per ton.[33] The most significant profits therefore arose from the sale of high-priced household coal because, as Buddle reiterated, "Every chaldron of small coal produced at the top costs . . . as much as a chaldron of the best."[34]

The irreducibility of high working costs was a constant refrain among coal owners. In 1829, Buddle was led to suggest that under competitive conditions, the only area in which coal concerns could reduce costs effectively was labor. However, there was a veiled threat in his conclusion announced before the parliamentary Select Committee that the reduction of wages could not "be effected without danger to the tranquility of the district."[35] It was, moreover, commonly asserted and privately accepted that competitive pricing inevitably forced profits below costs. Buddle laid out the following figures for Lord Londonderry in 1826.[36]

128,132 chs sold for		£157,146
average p.ch.		24/6
Cost working		126,000
average p.ch.		19/8
Profit	4/10	£31,146

If increase of production to 145,000 without vend and at reduced price:

145,000 chs at 20/6	£148,625
Cost of working	− 142,734
Apparent Profit	£ 5,891

with 57,000 ch. chargeable with tentale rent
(3,166½ tons at 30/. − £4,749.15.0)

---

[32] *Select Committee of the House of Lords on the State of the Coal Trade*, P.P. (1830), vol. 8, 66.

[33] D.C.R.O., Londonderry Papers, D/Lo/E 514(8), Abstract of Vend of Coals from 1838–1857.

[34] *Select Committee on the State of the Coal Trade*, P.P. (1830), vol. 8, 282.

[35] *Select Committee of the House of Lords on the State of the Coal Trade*, P.P. (1830), vol. 8, 68–9.

[36] D.C.R.O., Londonderry Papers, D/Lo/B 310(5).

Apparent Profit	£5,891
Deduct D + C Tentale	
rent	−4,749.5
Profit	£1,141.5

According to these estimates, despite the fact that the volume of sales might increase by more than 13 percent in open competition, the margin of profit per chaldron would drop from nearly 20 percent to only 4 percent while working costs (mostly labor) would rise 13 percent. All in all, Buddle estimated, on an increased volume of sales of nearly 450,000 tons, profits after additional rents would fall from over £31,000 to just over £1,100.

The future Lord Durham emphatically explained to Lord Londonderry upon the latter's entrance into the coal trade that open competition and free trade resulted in substantial financial losses. "That which I wish most strongly to impress on your mind," he wrote,

> is, that a *reduction of those high prices will inevitably follow a non-regulated Trade.* Yours will always be higher than the inferior coals, as it now is: but it will also be proportionately reduced with them. For instance – this year our Wallsend Coals have averaged 49ˢ the chaldron in London – next year if the regulation is broken a reduction of prices will take place throughout the whole list, to the amount of 3, perhaps 4, shillings a chaldron – the Wallsend will then be sold at 43 or 44 shillings, which on 100,000 chaldrons will make a *loss to the Proprietor* of £15,000 or £20,000 per ann. – I say to the Proprietor, for on him the reduction will fall.[37]

The acute awareness of ownership's limited control over costs affected industrial policy not only by facilitating cartellization and hence relocating the locus of control to the market, but also by encouraging the adoption of technology that raised the market price of commodities rather than directly reducing unit costs. Flinn has admirably surveyed vast areas of technological improvements in drainage, haulage, lighting, and ventilation that mitigated the "general and remorseless tendency within an extractive industry towards diminishing returns and rising marginal costs."[38] However, he failed to note that many of these developments were not viewed as a competitive edge on a larger volume of output, but instead as a means toward raising price against market pressures.

The introduction of "spouts" was one such innovation. Spouts were staithes equipped with chutes that deposited coal directly from the colliery wagon into the ship's hold. They were, necessarily, built below the Tyne Bridge at Newcastle or in the deeper sections of the Wear and thus obviated

---

[37] Lambton Mss., Lambton to Londonderry, 22 November 1823 (copy letter).
[38] Flinn, *History of the British Coal Industry,* 311, 442–8; see also Mitchell, *Economic Development,* 318.

the employment of keelmen and casters who previously had been employed to transport coal on barges to the deeper reaches of the rivers.[39] The keelmen's rich corporate history has been examined by Fewster and Rowe, both of whom have noted the marked change in the objectives of keelmen's strikes from limiting competition by controlling overweight measures to attacking directly the introduction of spouts.[40] Thus, for example, in 1816, Thomas Fenwick, the manager of the Church's mining interests, reported that the attempt by Nesham & Co. to extend the workings at Newbottle Colliery were seriously limited by the keelman's riot at Sunderland, which had destroyed the spouts and machinery there valued at £4,000.[41]

While cost reductions may seem to have been the logical effect of the destruction of the keelmen's trade,[42] the introduction of spouts, it was argued at the time, served to reduce handling and breakage and thus raise price. (The size of coal was an important determinant of price. Large, so-called round coals were much preferred by consumers.) Edward Steel and John Wood, two viewers who surveyed the Vane-Tempest estates in 1815, estimated that total profits there would increase by nearly £22,700 by shipping coals via spouts from Sunderland.[43] The viewers suggested that building a rail line and spouts would actually increase unit costs. Thus, by employing steam engines on cast iron rails, from Rainton Colliery to Sunderland, costs would be increased by about 1s. 1d. per ton. However, the reduced handling and breakage would increase the market price of the coal by 3s. 2½d. per ton, leaving an increased profit of 2s. 1½d. per ton. On an annual volume of 132,500 tons (or 50,000 chaldrons), profits were expected to rise an additional £13,854. When the alterations were applied to another Vane-Tempest colliery, Pensher, the price per ton was expected to increase by about 2s. 3d., which on 79,500 tons (30,000 chaldrons) would yield an additional £8,843 per annum. Steel and Wood thus had very little interest in the potential savings that might accrue on an increased volume of sales at lower costs and price, and given the marked inelasticity of demand, their considerations may not have been inappropriate.[44] Their eyes were set solely

---

[39] Ibid., 169–71.

[40] J. M. Fewster, "The Keelmen of Tyneside in the Eighteenth Century," parts 1–3, *Durham University Journal*, n.s., 14 (1957–8), 24–33, 66–75, 111–23; D. J. Rowe, "The Strikes of the Tyneside Keelmen in 1809 and 1819," *International Review of Social History*, 13 (1968); and idem, "The Decline of the Tyneside Keelmen in the Nineteenth Century," *Northern History*, 4 (1969).

[41] Records of the Palatinate of Durham and Bishopric Estates, Report Volumes, 1805–24, 93–4, in Department of Paleography & Diplomatic, Durham University.

[42] Flinn, *History of the British Coal Industry*, 168.

[43] D.C.R.O., Londonderry Papers, D/Lo/B 37(2).

[44] On the elasticity of demand, see Sweezy, *Monopoly and Competition*, and W. J. Hausman, "Market Powers in the London Coal Trade: The Limitation of the Vend, 1770–1845," *Explorations in Economic History*, 21 (1984), 389.

on the goal of raising profits by increasing the discrepancy between costs and price, particularly by raising the latter.

Another method of raising price by limiting losses caused by breakage, storage, and handling was called "tubbing." Tubs were fitted into coal wagons and could be hoisted directly by crane into keels.[45] This eliminated the need to "cast" coal from the keels into the colliers' holds and thus reduced the work of a specialized group of "casters."[46] However, we need not give credit to John Buddle for this development, as do Flinn and Hiskey. Tubbing seems to have been developed by William Croudace, J. G. Lambton's colliery agent in 1818.[47] As late as 1820, Buddle referred to the hostility aroused by "Mr. Lambton's mode of shipping Walls-end Coals by *Tubs*, which deprives the *Casters* (a turbulent Body of men) of employment." He further remarked that Lord Londonderry, Buddle's employer, would suffer the same unpopularity when and if he began shipping coal directly from Sunderland.[48] Nonetheless, Buddle hoped that the introduction of tubbing would increase the price of coals by an average of 2s. per chaldron.[49]

Inclement weather was a further factor that often played havoc with coal owners' revenues, and their response to seasonal disruptions again reveals the significance of price over markets. Northeasterlies could tie up shipping for days at a time, and winter freezes and spring floods seriously disrupted transport.[50] When the Wear froze in January 1823, Buddle explained to Lord Londonderry that "this has entirely sealed us up and put a stop to everything but our expenses, which *march on* steadily in spite of wind or weather."[51] Furthermore, the disruption of the flow of goods inevitably reduced the price of coal. Accumulating coal at the pit or storing them at the staithes broke up the large coals and reduced their quality and hence their price. Buddle estimated that storage reduced the value of the best coal from six to eight shillings a chaldron, which could amount to a loss of £50 a day.[52] It was, in part, this emphasis upon raising price that encouraged Lord Londonderry to undertake the construction of Seaham Harbor.

---

[45] Flinn, *History of the British Coal Industry*, 170.

[46] Flinn errs at least in part in assuming that keelmen were responsible for "casting" coals into the colliers. On the Wear, this was done by "casters."

[47] William Green, Jr., "The Chronicles and Records of the Northern Coal Trade in the Counties of Durham and Northumberland," *Transactions of the North of England Institute of Mining Engineers*, 15 (1866), 226.

[48] D.C.R.O., Londonderry Papers, D/Lo/C 142(143), Buddle to Londonderry, 15 April 1820.

[49] Sturgess, *Aristocrat in Business*, 34.

[50] For example, in July 1825, an ill wind tied up shipping in Sunderland harbor for several days. D.C.R.O., Londonderry Papers, D/Lo/C 142(148), Buddle to Londonderry, 26 July 1825.

[51] Ibid., D/Lo/C 142(144), Buddle to Londonderry, 18 January 1823.

[52] Ibid., D/Lo/C 142(148), Buddle to Londonderry, 3 July 1825.

Finally, the "screening" of coals was perhaps the most obvious and prevalent means of raising price. Screens, as the name implies, were bars spaced apart at certain distances so that only coals of a particular size passed through and separated out "small" coal from round coal. Almost all collieries screened their coal, and in the late 1820s, an average of over 25 percent of the coal hewn below ground in Tyneside collieries was screened out.[53] Screening created a more uniform commodity that could sell at higher prices. Buddle referred to the product as "manufactured coal" because it was not sent to market "in the state the mine produces it."[54] The net result was recognized as "an enormous waste" but one that raised the price of the proportion of coals that went to market.[55]

It certainly would be unwise to suggest that efforts to raise market price constituted the only impetus, albeit sometimes unintended, toward rationalization.[56] Leading, or the transport and haulage of coal, was one area in which the reduction of unit costs was perceived to be of great benefit. Thus, Arthur Mowbray advised the executors of the Vane-Tempest estates that "the use of steam has been in many instances successfully applied to waggons on the railways in the place of horses which makes the expense of leading, as to distance, of considerable less moment."[57] However, the historian risks inferring a degree of economic rationality that was neither pervasive nor wholly accepted among the coal owners of the early nineteenth century if the signal importance of raising price is ignored. In fact, attempts to raise price were in part a legacy of the "corporate sense" that reflected an unwillingess to accept fully the dictates of commercial society. Moreover, they were a pronouncement of the industry's struggle to control the market in preference to seeking control of the production process.

Competition and collusion

Controlling the market, therefore, rather than expanding the volume of sales and reducing unit costs, remained one of the principal objectives of coal owners before midcentury.[58] And nowhere can this be seen more clearly

---

[53] Flinn, *History of the British Coal Industry*, 106–8.
[54] *Select Committee on the State of the Coal Trade into the Port of London*, P.P. (1836), vol. 11, 48.
[55] Ibid.
[56] Cartellization could indeed encourage rationalization and innovation because, as Jürgen Kocka has pointed out, when "prices were fixed this was sometimes the only remaining method of increasing profits." See J. Kocka, "Entrepreneurs and Managers in German Industrialization," in P. Mathias and M. M. Postan (eds.), *The Cambridge Economic History of Europe: The Industrial Economies: Capital, Labour, and Enterprise*, vol. 7: pt. 1 (Cambridge University Press, 1978), 563–4.
[57] D.C.R.O., Londonderry Papers, D/Lo/B 36–7 (2).
[58] It has recently been argued by Hausman that "monopoly" profits were, in fact, quite modest,

than in the policies of the coal cartel. Buddle put the matter most succinctly: "The sole object of a regulation is to *gain price*."[59] However, the coal owners' dedication to supporting prices was often compromised either by their desire to secure a larger quota for their own firms, as Sweezy emphasized,[60] or by their efforts to obstruct the entry of new competitors into the trade.[61] In both cases, the largest firms that shipped their coals on the River Wear took the lead in establishing industrial policy. These firms, particularly those owned by Lord Londonderry and Lord Durham, showed no singular dedication either to the principles of the cartel or to a policy of free trade. Both the cartel and free trade were seen as tools by which owners' control of the market and competition might be established. Their source of strength was the productive capacity of their firms: They alone possessed the power to bludgeon the trade into submission by forcing a period of free, or "fighting," trade. Market competition, therefore, was neither a philosophy of bourgeois hegemony nor a pathway of freedom and enterprise. Instead, free trade was a weapon in the hands of the great coal owners of the early nineteenth century employed to secure their market preeminence.

The example of the entrance of the powerful Hetton Coal Company into the cartel in 1825 reveals not only how difficult it was for major new firms to enter the trade, but also the manner in which commodity markets were manipulated in order that the established leaders of the trade could maintain their superiority. The Hetton Coal Company first began shipping coals on the Wear in November 1822.[62] In March of the following year, the company initiated negotiations to join the cartel, itself due to be renewed at the end of the month.[63] By July 1823, no amicable agreement had yet been reached. At a meeting of the Wear coal owners in Chester-le-Street, County Durham on 11 July, the Hetton Coal Company's representatives, Nathaniel Scrutor, Matthias Dunn, and Arthur Mowbray, demanded an allotment of the trade equal to that of the dominant concerns of the Lambtons and Londonderry.[64] At the next meeting of the cartel on 21 July, Lambton (later Lord Durham) personally insisted on keeping Hetton to a maximum of one-half the pro-

---

perhaps "a couple of shillings per chaldron." But this may have had little effect on the perceptions of the trade at the time. See Hausman, "Market Power in the London Coal Trade," 383–405.

[59] D.C.R.O., Londonderry Papers, D/Lo/C 142(144), Buddle to Londonderry, 27 July 1823.

[60] Sweezy, *Monopoly and Competition*, 115–16.

[61] On industrial policy, see also Christine E. Hiskey, "The Third Marquess of Londonderry and the Regulation of the Coal Trade: The Case Re-opened," *Durham University Journal*, n.s., 44, no. 2 (1983), 1–9.

[62] R. Galloway, *Annals of Coal Mining and the Coal Trade* (London, 1898; repr. Newton Abbot, 1971), 450.

[63] D.C.R.O., Londonderry Papers, D/Lo/C 142(144), Buddle to Londonderry, 13 March 1823.

[64] Ibid., 15 July 1823.

portion of his own quotas. He also made it clear that he was willing to risk a period of intense competition and low prices in order to drive the Hetton Coal Company out of the trade. Lambton was reported to have said that there was no sacrifice he was not prepared to make. He let it be known that he "would put down the price of his Coals and run the race of destruction with [Hetton] 'till they were ruined."[65] John Buddle advised Lord Londonderry at the same time "that my private feeling is for war. The bubble of a Regulation must burst sooner or later as it has often done before."[66]

By early August 1823, the threat of free trade against powers of far greater capacity and capital had forced the Hetton Coal Company to acquiesce in the demands of Lambton.[67] Both Lambton and Londonderry were well satisfied that they had succeeded in maintaining their dominance in the trade. On 11 August 1823, Londonderry explained to Lambton that securing the mutual superiority of their concerns was his paramount interest: "Lord Londonderry will always continue desirous to cooperate with Mr. Lambton on all arrangements connected with their joint interests in the Coal Trade on the Wear, and more especially to preserve that superiority which their possessions entitle them to, against all Innovators and speculating adventurers."[68]

Nevertheless, the entry of the Hetton Coal Company into the cartel appears to have disrupted the Limitation of the Vend. By July 1824, the cartel had collapsed, free trade reigned, and Buddle blamed the Hetton Coal Company for selling more than their quota.[69] The repeal of the Combination Acts in 1825 and the subsequent union activity among several groups of local workers flung the trade into further disarray. The Hetton Coal Company, however, now used this situation to flex their industrial muscle. In March 1825, reports circulated throughout the North that Hetton had hired one hundred additional miners and was in the process of sinking two new pits. It was clear to Buddle, at least, that the company intended "to work every chaldron of Coals they can possibly raise; and they will sell them too, at any price they can get."[70]

In April 1825, Arthur Mowbray, Hetton's principal representative to the coal trade meetings at Chester-le-Street, resubmitted his demand for an allotment equal to that of Lambton's and Londonderry's concerns.[71] Tense

[65] Ibid., 21 July 1823.
[66] Ibid., 27 July 1823.
[67] Ibid., 8 August 1823.
[68] Lambton Mss., Londonderry to Lambton, 11 August 1823.
[69] D.C.R.O., Londonderry Papers, D/Lo/C 142(147), Buddle to Londonderry, no date (c. March 1825).
[70] Ibid., 23 March 1825.
[71] Ibid., 6 April 1825.

negotiations continued through April, May, and June of 1825. Mowbray apparently indicated that the Hetton Coal Company might be willing to accept allotments at a proportion of three-fourths that of the Lambton and Londonderry quotas, but together those proprietors insisted on keeping them to one-half the proportion of the largest concerns.[72] By mid-June, Londonderry's resolve began to weaken. The banking crisis of 1825 had taken a severe toll on his personal and industrial finances, and he began to search for a solution that would end open competition.[73] Lambton, however, refused to yield to the so-called Hettonians. Lord Londonderry tried to effect a settlement, but he wrote to Lambton that the negotiations made him feel as if "we are Crouching to this power beyond all present Justice and fair Dealing."[74] On 24 June, a private meeting in Newcastle between Lambton and representatives of the Hetton Coal Company failed to conciliate the parties. Buddle became increasingly dismayed as the industrial might of the Hetton Company continued to be revealed. "The more I learn of the views of the Hetton Co.," he wrote Lord Londonderry, "the more I am convinced that a great Shock is approaching. My firm opinion is come it must, as no possible increase of demand can meet the increased powers of supply which are now in preparation."[75]

Suddenly, on 30 July, a new cartel agreement, including the Hetton Coal Company, was signed whereby Hetton was given an allotment equal to three-fourths that of Lord Londonderry and Lambton.[76] Although it may be tempting to describe the movement to readopt the regulation as the expected response of capitalists striving to restore monopoly profits, this was not the case.[77] Instead, the coal owners of the Wear reunited only in order to better defend themselves against trade union agitation. The last weeks of July 1825 were marked by the beginning of widespread union activity in the Northeast. According to Buddle, a miners' union already had been completed on the Tyne by 12 July.[78] Moreover, the Seaman's Loyal Standard Association was actively recruiting members at the same time.[79] On 23 July, Buddle wrote that "the menacing aspect of the Pitmen has had some affect I think in disposing the Coal-owners to be a little more unanimous than they might

[72] Ibid., 15 June 1825.
[73] Ibid., 13 June 1825.
[74] Lambton Mss., Londonderry to Lambton, 14 June 1825.
[75] D.C.R.O., Londonderry Papers, D/Lo/C 142(148), Buddle to Londonderry, 2 July 1825.
[76] Ibid., 30 July 1825.
[77] Joe S. Bain, *Barriers to New Competition* (Cambridge, Mass.: Harvard University Press, 1956), 36.
[78] D.C.R.O., Londonderry Papers, D/Lo/C 142(148), Buddle to Londonderry, 12 July 1825.
[79] See D. J. Rowe, "A Trade Union of the North-East Seamen in 1825," *Economic History Review*, 2nd ser., 25, no. 1 (1972), 81–98.

otherwise have been."[80] Two days later, a detachment of troops was called in to Sunderland in order to protect ships and nonunion sailors while another detachment was requested to be stationed at South Shields before the end of the week.[81] Only then, on 30 July, was the cartel agreement renewed. The case of the Hetton Coal Company reveals the extent to which the great collieries of the Wear were determined to maintain their industrial superiority. Free trade and open competition was the bludgeon used to maintain industrial discipline and to secure the dominance of the largest firms.

Open competition could perform other instrumental functions within the trade as well. For example, free trade often lured industrialists who envisioned getting the jump on their competitors by vending large quantities of coal before prices at market fell. This presented a particularly tempting prospect when incomes were strained and quick infusions of cash were needed. "The point to be aimed at," Buddle explained to Lord Londonderry, "is, to make the greatest possible Vend, at the least possible reduction in price."[82] Similarly, part of J. G. Lambton's rationale for breaking with the cartel in 1829 was that his spendable income had fallen dramatically. It is no doubt significant that it was not Lambton's colliery agent, Henry Morton, who convinced him to plunge into a fighting trade, but his estate auditor, H. F. Stephenson.

In January 1829, Stephenson urged Lambton to declare his independence from the cartel because the payments on personal debts had risen unexpectedly at the same time that receipts from the landed estates had declined dramatically. Thus, while colliery profits in 1828 had been strong, they totaled £21,833 in that year, net payments on all personal accounts had risen to over £31,000. Based on these figures, Stephenson wrote to Lambton, "Your spendible income was no more than £914."[83] After reviewing past colliery sales records, Stephenson advised that an open competition in the coal trade was the quickest path to financial recovery. The auditor argued that during the last period of free trade in 1825–6, Lambton's total coal sales had reached £191,000 and his prestigious Wallsend coals had been in great demand. At that time, although total colliery profits had dipped to about only £6,400, an additional £60,000 had been diverted to the winning of Houghton Pit.[84] Thus, Lambton was convinced that the revival of his firm's profits lay along the route of open competition.

[80] D.C.R.O., Londonderry Papers, D/Lo/C 142(148), Buddle to Londonderry, 23 July 1825.
[81] P.R.O., Kew, H.O., 40/18/284; ibid., 40/18/288.
[82] D.C.R.O., Londonderry Papers, D/Lo/C 142(149), Buddle to Londonderry, 16 April 1826.
[83] Lambton Mss., H. F. Stephenson to Lord Durham, 3 January 1829.
[84] Lambton Mss., Stephenson to Durham, 13 January 1829.

Despite Stephenson's recommendations, an open coal trade could not be sustained profitably for long. By September 1829, the cartel had been re-assembled, and Lord Durham wrote with relief to the third Earl Grey that "we have at last effected a regulation after having in 8 months lost on Tyne & Wear nearly £500,000; in another month we must have shut up the mines."[85] The short-term prospect of a quick return had developed into a contest of attrition. As Buddle had counseled Lord Londonderry during an earlier bout of unrestricted competition, "Free Trade so far, has certainly been our best policy, but . . . finding out the proper point at which to stop is the difficulty."[86]

Before 1840, therefore, neither the principles nor the discourse of classical political economy held complete sway over the coal trade. Instead, the corporate legacy of the preindustrial era continued to exert a profound influence upon both ideology and policy. The great coal industrialists before mid-century more closely resembled those Continental entrepreneurs who, according to Landes, exhibited a pronounced "preference for the greatest possible profit per unit of sale, as against higher total profit at some larger output" and who condemned competition as both unfair and subversive.[87]

The coal owners' perception of the market was essentially adversarial and instrumental. The industry publically argued that the laws of supply and demand not only were inapplicable to the northern coal trade, but they also were counterproductive and against the best interest of the consumer, Among themselves, free trade was a weapon employed in the industrial conflicts generated by capital shortages and contests for industrial dominance. In sum, the market was viewed as a legitimate mechanism through which resources were allocated and prices established only when it applied to wages; there was to be no commodity market, as it was defined by political economists. A semblance of the acceptance of market principles took hold only after 1844, the date of Lord Londonderry's defection from the cartel, when the presence of expanding local industrial and foreign markets absorbed the lower-quality and lower-priced small coal that had burdened the industry for decades.[88]

[85] Grey Papers, Durham University, Department of Paleography Diplomatic, Durham to Grey, 4 September 1829.

[86] D.C.R.O., Londonderry Papers, D/Lo/C 142(149), Buddle to Londonderry, 24 June 1826.

[87] Landes, *Unbound Prometheus*, 132–3. However, unlike their Continental counterparts described by Landes, the subversion of price mechanisms in the northern coal industry did not inhibit rationalization.

[88] See James A. Jaffe, "Economy and Community in Industrializing England: The Durham Mining Region before 1840," (Ph.D. dissertation, Columbia University, 1984), 65–8. On the end of the regulation, see David Large, "The Third Marquess of Londonderry and the End of the Regulation, 1844–5," *Durham University Journal*, n.s., 20, no. 1 (1958), 1–9; Hiskey, "The Third Marquess of Londonderry," 1–9.

3

Managerial capitalism

The owners' efforts to control the market were not located solely in the realm in which commodities circulated. The other principal area in which they sought to establish control was in the labor market. This struggle to define the terms of market, as opposed to production, relations was located both at the workplace and within the community at large. It was in these areas that the struggles to control exchange relationships were realized as class struggles.

The struggle to control the terms of market relations with labor were all the more important for management and owners since they accepted the relative autonomy of the worker at the point of production. This might be considered particularly surprising since the northern coal industry was mid-wife to the birth of modern managerialism during the Industrial Revolution. As the previous chapters have indicated, the mines of the Northeast required extensive accumulations of capital both to start up and to maintain production. Indeed, the largest firms of the Northeast were among the most heavily capitalized firms in the nation. The expansion of production underground and the employment of expensive capital equipment brought in its train the specialization of managerial functions in the hands of professional "viewers." The significance of the viewer in the history of British economic development can hardly be overstated. They were, as Sidney Pollard noted, the direct ancestors of the great railway builders of the nineteenth century.[1] Moreover, their influence not only was regional but extended throughout the nation and to almost all areas of engineering.[2]

Nevertheless, the development of a professional managerial stratum within firms did not engender either a drive for control at the point of production or a concomitant effort to deskill the worker. While it would be naive to assume that industrial relations in the northern coal industry as a result were conciliatory and quiescent (quite the opposite; P. K. Edwards's term "structured antagonism" more accurately reflects the underlying conflict between

[1] Pollard, *Genesis of Modern Management*, 126–8.
[2] Flinn, *History of the British Coal Industry*, 63–8.

workers, management, and owners),[3] shopfloor relations were negotiated to a surprising extent. This was due largely to the fact that management was unwilling and, given the ever-changing geography of production, unable to exercise direct control at the coal face.

Instead, managers relied upon the frequent revision of piece rates that took place through a recognized, albeit informal, process of workplace bargaining. The key to management's power within the production process was the inspection of the finished product and the imposition of fines for poor workmanship, both of which were elaborated in the annual terms of employment, the miners' bond. Therefore, the viewers' role in the development of modern managerial practices may be considered a more limited one than previously acknowledged. While they pioneered the separation of ownership from management and were in the forefront of applying scientific and engineering advances to industry, their relationship to labor within the production process was more akin to the factor of the "putting-out" system than to the scientific manager of the late nineteenth and early twentieth centuries. Their "formal" control of the labor process was much more apparent than their "real" control.[4]

Management's inability to control the point of production, however, did not mean that it failed to construct effective authority over the work force. On the contrary, the locus of managerial control tended to shift outside of the workplace and into the community, where company-owned housing predominated. In this way, if management proved unable to control the labor process, it could at least attempt to control the labor market.

Of equally profound significance, was the effect of the separation of ownership from management on industrial relations in the coal industry. Labor agitation and social conflict in the nineteenth century focused on the relationship between viewers and workers, not owners and workers. Perhaps surprisingly, this enabled owners to retain the paternalistic loyalty of their workers even in the midst of widespread social unrest. Thus, the social position of ownership was secured as labor grievances were directed toward management and not property owners.

Management, therefore, was one of the principal actors in the construction of market relations in the early-nineteenth-century coal industry. Its social power and influence rested upon its role as frequent negotiator and interpreter of the annual Vend agreements between coal owners as well as the annual bond agreements with labor. While the Vend operated to subvert the ap-

---

[3] P. K. Edwards, *Conflict at Work: A Materialist Analysis of Workplace Relations* (Oxford: Blackwell Publisher, 1986).
[4] On this distinction, see Gareth Stedman Jones, "Class struggle and the Industrial Revolution,"*New Left Review*, no. 90 (March–April 1975) reprinted in idem, *Languages of Class*.

plication of market forces to commodities, management sought to influence the application of market forces to labor through the terms of the bond and the control of housing. Toward this end, managerial power could be and was employed despotically, most frequently in regard to the imposition of fines for poor workmanship. However, the persistence of piece work coupled with the lack of managerial oversight entailed a negotiated relationship with labor that was epitomized in the practice of workplace bargaining.

Correspondingly, management's inability to control the labor process at the point of production made their control of the commodity and labor markets all the more important. Therefore, throughout the first half of the nineteenth century, class conflict in the northern coalfield focused upon the mastery of these markets or, more specifically, the definition of the terms of exchange within the market, rather than upon the control or appropriation of the means of production.

Ownership

The relationship of landed society and capital to industrial development is nowhere more important than in the coalfields of Durham and Northumberland. There is general agreement that before the nineteenth century, the aristocracy and gentry of the North played the principal role in the development of the regional coal industry.[5] John Bailey's agricultural survey of Durham published in 1810 listed thirty-four collieries in the county that shipped coal outside of the immediate region. (These are the so-called watersale collieries. They are distinct from landsale collieries, which produce for local consumption alone. Watersale collieries invariably employed more capital and labor; they also comprised the coal owners' cartel.) Just less than one-third of these watersale collieries were owned and operated by members of the local landed elite. These owner-operators included one earl, four baronets, and two squires. One of these squires, in fact, was W. H. Lambton, a political confidante of Charles James Fox who took a bit of pride in his family's nonnoble status as well as in the fact that his father had refused the offer of a peerage.[6] His son, J. G. Lambton, or "Radical Jack," would later accept ennoblement and then be raised to an earl in lieu of a position of influence after the Reform Bill crisis. The Lambtons' gentry status in Durham was hence purely a matter of form and not substance.

About another third of the county's collieries were owned wholly or in

---

[5] Flinn, *History of the British Coal Industry*, 36–40; David Spring, "English Landowners and Nineteenth Century Industrialism," in Ward and Wilson, *Land and Industry*.

[6] Grey Papers, Durham University, Department of Paleography Diplomatic, Lambton to Grey, 23 August 1831.

part by the bishop of Durham or the cathedral's dean and chapter. By the nineteenth century, the Church invariably leased out its mining properties to speculators and entrepreneurs. However, leaseholders of the Church's properties still were dominated by the gentry and aristocracy. The county's great landed families were numbered among the lessees of the Church's ten watersale collieries, including the Lambtons, Liddells, Neshams, Brandlings, and Vane-Tempests.

The remaining collieries in the county were owned overwhelmingly by members of the landed class, but these mines were not worked directly by their proprietors. Only three commoners were proprietors of the remaining fourteen watersale collieries listed by Bailey. Yet even when the noble proprietors chose to lease out their estates rather than work the mines themselves, commoners rarely appeared among the entrepreneurs. The Earl of Scarborough, for example, leased his Lumley Colliery to W. H. Lambton. Sir John Eden, Bart., owner-operator of Beamish Colliery in 1810, let out Herrington Mill Colliery to Sir Henry Vane-Tempest. Similarly, Sir Thomas Liddell, Bart., owner-operator of Low Moor and South Moor collieries, leased Ravensworth Colliery to George Burdon, Esq.[7]

The beginning of the nineteenth century, therefore, witnessed a virtual aristocratic monopoly not only of the ownership of land, as one would expect, but also of the exploitation of coal. It has been argued by some, however, that the nineteenth century was marked both by the removal of the upper classes from active enterprise and by their gradual conversion to the position of a rentier aristocracy. Eric Richards, for one, has argued that in the case of the Stafford family, their aristocratic ideals were eventually confronted by the social effects of industrialization. Given the conflict between deference and development, the Staffords opted for the former.[8] Others have argued that social and ideological rationales were not the only reasons for the growing tendency toward a rentier aristocracy in the nineteenth century. In the specific case of mining, J. T. Ward claims that the growing scale of capital requirements, the consequent increase of financial risk, and the increasingly technical nature of extractive industry all militated against the continued involvement of the aristocracy.[9]

---

[7] John Bailey, *General View of the Agriculture of the County of Durham* (London, 1810), 12–21. A similar point is made by Flinn, *History of the British Coal Industry*, 38–9.

[8] E. Richards, "The Industrial Face of a Great Estate: Trentham and Lilleshall, 1780–1860," *Economic History Review*, 2nd ser., 27, no. 3 (1974). See also, François Crouzet, *The First Industrialists: The Problem of Origins* (Cambridge University Press, 1985), 79–81. The *locus classicus* of this view is expressed in Martin Wiener, *English Culture and the Decline of the Industrial Spirit, 1850–1980* (Cambridge University Press, 1981).

[9] J. T. Ward, "Landowners and Mining," in Ward and Wilson, *Land and Industry*, 72–5. See also Sill, "Landownership and Industry," 146–66.

Neither of these arguments, however, adequately applies to the northern coalfield. Flinn has rightly stressed the fact that the interlocking patterns of leaseholding and the aristocratic interest in the development of their coal mining properties combined to secure direct elite involvement in industry throughout the first third of the nineteenth century.[10] Indeed, a list of the major coal proprietors as late as 1890 would look very little different than it had a century earlier. Three of the five largest mining firms in Durham in 1893 were still those of the earl of Durham, the marquess of Londonderry, and John Bowes.[11]

While elite involvment in the northern coal industry did not necessarily decline absolutely, aristocratic and gentry entrepreneurs did begin to play a smaller role relative to the more quickly growing joint-stock companies and partnerships. For example, the two decades after 1820 witnessed the rapid expansion of mining into east Durham, where the coal seams lay below a massive strata of magnesian limestone. The Hetton Coal Company, founded in 1820, was the first enterprise to pierce this layer and open up the coalfield. Originally composed of eleven shareholders, the capital for the undertaking came mostly from a collection of local attorneys, viewers, and fitters.[12] The eventual success of Hetton, not only in winning coal from beneath the magnesian limestone but also in breaking into the coal cartel, acted as a lure to other investors.[13] Already in 1829, John Buddle noted that owner-operated firms were rare on both the Tyne and Wear. He cited only eight of fifty-nine collieries on the two rivers that were operated by their landed proprietors.[14] By 1840, the Haswell Colliery Company, the South Hetton Colliery Company, the Thornley Colliery Company, and the Wingate Grange Colliery Company all were active in the east Durham coalfield.[15]

For their part, the elite coal owners viewed the proliferation of coal mining companies with fearful resignation. On the one hand, it was not unusual for members of the Durham nobility to continue to enter into partnerships to augment their investments. Thus the Liddells, Lambtons, and Lord Strathmore all held interests in collieries other than their own.[16] On the other hand, these landed families feared the ability of the new companies to raise capital and to compete effectively. In 1830, Lord Londonderry warned Lord

[10] Flinn, *History of the British Coal Industry*, 38.
[11] D.C.R.O., Londonderry Papers, D/Lo/B 11(13).
[12] Ibid., D/Lo/B 309(19); Sill, "Landownership and Industry," 151–5; Flinn, *History of the British Coal Industry*, 209.
[13] On Hetton's stormy entrance into the cartel, see Chapter 2.
[14] *Select Committee on the State of the Coal Trade*, P. P. (1830), vol. 8, 31, 270.
[15] Sill, "Landownership and Industry," 158; Galloway, *History of Coal Mining*, 203–10.
[16] T. Y. Hall, "Old Mining Records and Plans," *Transactions of the Institution of Mining Engineers*, 81 (1930–1), 80–1.

Durham that the limits placed on production by the cartel actually acted to spur speculation in the northern coal industry by creating a scarcity of coal. He wrote: "I confess I have very serious doubts of the policy of encouraging all these new Collieries, and we shall find too late as in the Hetton instance that the old ones are beaten out of the field. For individuals of Capital and *adventurers* can not compete with these cursed companies."[17]

As direct owner-operation declined in relative significance, leaseholding arrangements became increasingly important.[18] Since the Church rarely worked its own mines, its policies reveal many of the most significant trends in leasing during this period. When Parliament surveyed Church leases in the late 1830s, it found that Church property in County Durham had an annual total value of approximately £47,000.[19] Of this, mining accounted for over £18,000 annually, or about 39 percent. While the Church does not appear to have followed any systematic leasing policy, there was a general tendency for it to lease mining properties for years (twenty-one years was most common) and other properties for lives. Mining leases account for 57 percent of the annual value of all leases held for years, but only 21 percent of leases held for lives.

In order to encourage the development of their mining properties, the Church often kept annual rents at a relatively modest level but then imposed stiff renewal fines every seven years. Needless to say, this often embittered relations between some of the major coal owners and the Church, so much so that by the late eighteenth and early nineteenth centuries, it began to avoid this policy. Instead, the Church offered leases that were directly linked to the amount of coal produced by the colliery. This "tentale rent" levied an annual charge on a set quota of production. When the lessee exceeded that quota, a premium was exacted for each "ten" (17.5 chaldrons, or about forty-six tons) that was raised from the pit. In this manner, the Church not only secured a more regular flow of income but also was able to keep up with the expansion of coal production.

While most coal producers resented the old fining system, tentale rents threatened to increase substantially their burden of costs. For example, in 1814, the trustees of the Vane-Tempest estate were advised as to the comparable charges of leasing Rainton Colliery by a tentale rent rather than an annual rent and fine. Steel and Wood, the viewers who were hired to survey the mining properties of the estate for the trustees, estimated that the renewal

---

[17] Lambton Mss., Londonderry to Durham, 16 September 1830.
[18] See Flinn, *History of the British Coal Industry*, 43–9; Ashton and Sykes, *Coal Industry*, 175–181; David Spring, "The Earls of Durham and the Great Northern Coalfield, 1830–1880," *Canadian Historical Review*, 23, no. 3 (1952), 243; and Sturgess, *Aristocrat in Business*, 12.
[19] *Report from the Select Committee on Church Leases*, P. P. (1838), vol. 9, 525–6t.

of the Rainton lease by tentale rent of twenty shillings per ten – "a fair
Tentale Rent for this Colliery," they noted – would cost the estate an
additional £17,143 over the succeeding seven years compared to the previous
system of fines.[20]

Not all members of the Church, however, were convinced that tentale
rents served their best interests. In Durham, older members of the dean
and chapter fought against this change because tentale rents spread out
income over a long period. In 1820, when the Rainton lease came up for
renewal again, the notorious Henry Phillpotts, later bishop of Exeter and
scourge of Nonconformists and Tractarians alike, led a contingent of younger
"anti-Fineists" who hoped to alter the Rainton lease to a tentale rent. Older
members of the chapter fought Phillpotts so as to reap immediate benefits
from the large fine expected from the renewal of the lease. As John Buddle
remarked of one of Phillpotts's opponents, "He is wishing to sell his *Stall,*
[and] he is anxious to *sack all the Pewter* he can before he goes."[21] Buddle
advised John Iveson, agent for the Vane-Tempest estates, to ready himself
for battle. "As I expect shortly to see you in *the Ring* again with Phillpotts,"
he wrote, "I beg that you may put yourself into immediate *training....* If you
cannot get a better Man I will second you myself as well as I can, and I beg
of you to practice the *Chopper* diligently, so as to be able to put it in neatly,
as you know your man is *Shifty.*"[22]

Subcontracting

Subcontracting as a form of industrial organization and management occurred only briefly and under unique circumstances in the Northeast. This
system, under which proprietors contracted out the working of their mines,
was very common in the Midlands, where enterprise was on a much smaller
scale. As in other industries where subcontracting was common, it often led
to the accelerated exploitation of labor as contractors forced down wages
while speeding up production, as well as the uneconomical use of natural
and mechanical resources for immediate rather than long-term benefits.[23]
Although subcontracting certainly was not unknown in the northern coalfield

[20] D.C.R.O., Londonderry Papers, D/Lo/B 35.
[21] Ibid. D/Lo/C 142(143), Buddle to Iveson, n.d. (1820).
[22] Ibid.
[23] A. J. Taylor, "The Sub-contract System in the British Coal Industry," in L. S. Pressnell
(ed.), *Studies in the Industrial Revolution: Presented to T. S. Ashton* (London: Athlone, 1960),
218–9. For subcontracting in the building trades, see Richard Price, *Masters, Unions and
Men: Work Control in Building and the Rise of Labour, 1830–1914,* (Cambridge University
Press, 1980), 30–1.

– the Delavals had subcontracted Hartley Colliery as early as 1756 – it still was by no means common practice.

A. J. Taylor's influential article on this subject argued that there was a sudden proliferation of subcontracting in the Northeast between about 1780 and 1815. In 1784, the Lambtons first subcontracted their mines to Mssrs. Featherstonhaught & Co., and this was then followed by other Durham families: The Milbankes and Bowes had concluded similar arrangements by 1797, and the Tempest family contracted out their collieries in 1801. According to Taylor, the impetus to subcontract the working of collieries was the result of the inability of owner-operators to come to terms with the instability of the coal industry during the late eighteenth and early nineteenth centuries.[24] The American and Continental wars had led to a period of inflation, price fluctuations, labor shortages, union activity, and rising working costs. However, when the Napoleonic Wars ended, Taylor argued, prices steadied, labor was plentiful, and working costs fell. Thus, after 1815, there began a new trend back toward the direct control of collieries by their proprietors. Flinn thereafter argued that most coal owners found subcontracting "troublesome, expensive and . . . unproductive," as was reported to the Vane-Tempests in 1814.[25]

There does, however, seem to be more to the trends in subcontracting than economic rationality alone. Certainly in the case of the two largest coal owning families, the Vane-Tempests and Lambtons, the fortunes of family descent played a decisive role in the decision to subcontract. Upon John Tempest's death without a male heir, his estates were devised to Sir Henry Vane, Bart., in 1794. As part of the estate settlement, he assumed the name Vane-Tempest. Five years later, in 1799, he married Anne Catherine, countess of Antrim, coheiress to the Antrim estates in Ireland.[26] Sir Henry was renowned as a drunkard, gambler, and horse player; a mid-nineteenth-century Durham antiquarian thought him "in total want of intelligence," and this appears to have been a not uncommon notion.[27]

Sir Henry spent much of his time either in London or on the Continent, and the subcontracting of the estates in 1801 almost certainly occurred under these circumstances. The extraordinary nature of subcontracting the Vane-Tempest collieries is supported by the actions of the trustees of the estate after Sir Henry's death in 1813. Within a fortnight, still two years before the

[24] Taylor, "The Sub-contract System," 220–32.
[25] Quoted in Flinn, *History of the British Coal Industry*, 57.
[26] A. P. W. Malcomson, *The Pursuit of the Heiress: Aristocratic Marriage in Ireland, 1750–1820* (Belfast: Ulster Historical Foundation, 1982), 26–30.
[27] William Fordyce, *The History and Antiquities of the County Palatine of Durham* (Newcastle, 1857), vol. 2, 323.

end of the Napoleonic Wars, the trustees made a motion in Chancery on the part of Sir Henry's daughter, Lady (Frances Anne) Vane-Tempest, to appoint a manager of the collieries and to take back their direct control.[28] A subsequent survey of the estates decidedly advised against subcontracting.[29] In the end, after much litigation, the estates were put in the hands of Lord Stewart (later the third marquess of Londonderry) as life tenant upon his marriage to Lady Frances Anne in 1819.[30] Subcontracting on the Vane-Tempest estates is therefore more likely to have occurred in the context of personal and family concerns rather than as a response to fluctuations in the coal trade.

The circumstances of subcontracting the Lambton collieries are more problematic, but again appear to owe more to genealogy than economics. General John Lambton succeeded to the estates in 1761 at the age of fifty-one. His eldest son William Henry (b. 1764) attended Cambridge and in 1784 left on the Grand Tour. That same year, General Lambton subcontracted the collieries, not unreasonable for a man well past seventy whose eldest son is unavailable to help superintend the works. Shortly thereafter, however, the general's health began to fail, and he called his son back to England in 1786. Yet this did not solve the problem of managing the collieries. William Henry was himself in ill health. He spent many years in the Mediterranean hoping to cure his consumption, but he died at Pisa in 1796. This left five-year-old J. G. Lambton heir to the estates and the subcontractors in charge of the collieries. Immediately upon attaining his majority in 1813, J. G. Lambton resumed direct control of the collieries and dismissed the subcontractors.[31]

Subcontracting in the Northeast thus was a stopgap measure resorted to by unwilling or incapable heirs. Its appearance near the end of the eighteenth century was not necessarily due to cyclical variations in the coal industry, but partly to coincidental lapses in the descent of two great estates. Similarly, the connection between the decline of subcontracting in the northern coalfield and the end of the Napoleonic Wars is more apparent than real. In the case of the Lambton and Vane-Tempest estates, the resumption of direct exploitation of their collieries coincided with the descent of the properties to new heirs and trustees. And in both cases, this occurred well before the Continental Wars had been concluded.

---

[28] Ibid.

[29] D.C.R.O., Londonderry Papers, D/Lo/B 33, parts of which are quoted in C. E. Hiskey, "John Buddle (1773–1843), Agent and Entrepreneur in the North-east Coal Trade" (M.Litt. thesis, Durham University, 1978), and Flinn, *History*, 57.

[30] Sturgess, *Aristocrat in Business*, 6–10, 101–4; Malcomson, *Pursuit of the Heiress*, 30.

[31] Fordyce, *History and Antiquities*, vol. 1, 347–8; vol. 2, 630–1; *The Dictionary of National Biography*, vol. 17, 462–7; Stuart J. Reid, *Life and Letters of the First Earl of Durham 1792–1840*, 2 vols. (London: Longman Group, 1906), vol. 1, 16–66.

Management

It is generally true that during the early stages of industrialization, the functions of ownership and management were performed by entrepreneurs or their family members and trusted friends. "Owners managed and managers owned," was Alfred Chandler's succinct description of early industrial firms in America,[32] while Peter Payne described individual proprietorship and closely held partnerships as "the fundamental business unit of the industrial revolution."[33] However, the most significant exceptions to this generalization occurred in the management of both agricultural estates and mining concerns where professional estate managers and viewers were already being employed by the middle of the eighteenth century.[34]

Few viewers of coal mines seem to have had any professional training, in the modern sense. Similarly, apprenticeships were rare, although not unknown. According to Matthias Dunn, one of the early viewers of Hetton Colliery, the first viewer to serve an apprenticeship was Luke Curry around the year 1750.[35] Generally, training was acquired through personal observation, and positions were secured through personal recommendations. Not surprisingly, therefore, strong kinship ties existed among viewers: John Buddle was trained by his father; William Stobart and his two sons were viewers; Jonathan Walker trained his son, Wylam; and the grandfather, father, and three sons of the Smith family all were viewers in the North.[36]

---

[32] Alfred Chandler, *The Visible Hand: The Managerial Revolution in American Business* (Cambridge, Mass.: Harvard University Press, 1977), 9. For Britain, the activities of managers and entrepreneurs are followed most thoroughly in business histories such as Mathias, *The Brewing Industry in England, 1700–1830*, (Cambridge University Press, 1959); W. G. Rimmer, *Marshalls of Leeds: Flax-Spinners, 1788–1886* (Cambridge University Press, 1960); Stanley D. Chapman, *The Early Factory Masters* (Newton Abbott: David & Charles, 1967); Eric Sigsworth, *Black Dyke Mills* (Liverpool: Liverpool University Press, 1958); or, Neil McKendrick, "Josiah Wedgwood: An Eighteenth Century Entrepreneur in Salesmanship and Marketing Techniques," *Economic History Review*, 2nd ser., 12, no. 3 (1960), 408–33.

[33] Peter L. Payne, "Industrial Entrepreneurship and Management in Great Britain," in Mathias and Postan, *The Cambridge Economic History of Europe: The Industrial Economies*, 1, 92.

[34] G. E. Mingay, *English Landed Society in the Eighteenth Century* (London: Routledge & Kegan Paul, 1963), 173–7; F. M. L. Thompson, *English Landed Society in the Nineteenth Century* (London: Routledge & Kegan Paul, 1963), 151–83; David Spring, *The English Landed Estate in the Nineteenth Century: Its Administration* (Baltimore: Johns Hopkins University Press, 1963), 97–134; idem, "Agents to the Earls of Durham in the Nineteenth Century," *Durham University Journal*, 54, no. 3 (1962), 104–13; idem, "The Earls of Durham," 237–53; Pollard, *Genesis of Modern Management*; Flinn, *History of the British Coal Industry*, 49–68.

[35] Hall, "Old Mining Records and Plans," 77–9.

[36] Edward Hughes, "The Professions in the Eighteenth Century," 186, in Daniel Baugh (ed.), *Aristocratic Government and Society in Eighteenth-Century England* (New York: New Viewpoints, 1975); Hall, "Old Mining Records and Plans," 77–9.

The viewer's role in the coal industry often was an ambiguous one. In one sense, he was a "commissioned, salaried entrepreneur"[37] who was "in the forefront of the managerial revolution of the new industrial age."[38] As such, viewers were responsible for conducting labor relations, negotiating leases, purchasing equipment, colliery administration, and bookkeeping. Moreover, these vast administrative functions were surpassed by the viewer's engineering skills. He conceived and organized the construction of waggon ways and later railroads; he was among the earliest industrialists to employ steam engines; he planned the ventilation and structural support as well as the working out of the mines; and he organized and planned the construction of docks and staithes for the shipment of coal.

Moreover, these Promethean managers were often entrepreneurs in their own right. John Buddle was part owner of "several" collieries, he testified before parliamentary hearings.[39] Two viewers, William Stobart, Jr., and John Wood, were among the first partners in the Hetton Coal Company in 1820.[40] Viewers also were often hired as independent consultants by businesses to assess the value of a colliery's workings, to act as outside evaluators of a firm's operations, and to project the costs of modernization, repairs, or expansion. Thus, Steel and Wood were hired to advise the Vane-Tempest trustees as to the likely cost of modernizing those works in 1815, while Buddle, the most famous viewer of the the first half of the nineteenth century, consulted and advised coal owners not only in Durham and Northumberland, but throughout the country.[41]

This ambiguous position between manager and owner in the regional coal industry was compounded when viewers undertook to invest in individual firms. It is axiomatic both that coal owners provided company housing for the miners and that the provision of housing was an essential tool of control throughout the industry.[42] However, according to Henry Morton, the agent

---

[37] The phrase is Jürgen Kocka's used to refer to the German counterparts of Britain's viewers in "Entrepreneurs and Managers in German Industrialization," in Mathias and Postan, *The Cambridge Economic History of Europe: The Industrial Economies* 1, 493.

[38] Flinn, *History of the British Coal Industry*, 68.

[39] *Select Committee on the State of the Coal Trade*, P.P. (1830), vol. 8, 270. Flinn notes Buddle was part owner of five collieries over the course of his career: Flinn, *History of the British Coal Industry*, 59.

[40] D.C.R.O., Londonderry Papers, D/Lo/B 309(19); reprinted in Sill, "Landownership and Industry," 153.

[41] See D.C.R.O., Londonderry Papers, D/Lo/B 37 and 310, for Steel and Wood's report and Buddle's valuation of the Lambton collieries, respectively.

[42] Flinn, *History of the British Coal Industry*, 429–34; Rule, *Labouring Classes*, 99–101; Robert Colls, *The Pitmen of the Northern Coalfield: Work, Culture, and Protest, 1790–1850* (Manchester: Manchester University Press, 1987), 262–6; M. J. Daunton, "Miners' Houses: South Wales and the Great Northern Coalfield, 1880–1914," *International Review of Social History*, 25,

of the earl of Durham, up through the 1830s this was not always the case in Durham and Northumberland: "The Coalowners never build cottages for workmen almost," Morton wrote to Lord Durham's auditor in 1835. "The agents at Hetton," he said, "built a great part and those at So. Hetton were all erected on Speculation."[43] In January 1832, according to Matthias Dunn's diary, the proprietors of Hetton Colliery determined to expand their stock of company housing by delegating the job to speculators.[44]

The problem of providing company housing was brought home to Morton when he failed to secure financing for the erection of colliers' housing at Lord Durham's Sherburn Hill Colliery. In 1835, he had reached an agreement with Lord Durham to build 150 cottages and rent them out at £4 per annum for the next twenty-five years. However, the property, he only discovered later, was leasehold tenure of the bishop of Durham. Moreover, the deed stipulated that the housing and building materials on the property reverted to the bishop if they were not removed and sold when the lease ended. In effect, investments in housing could be totally lost. Therefore, Morton was unable to offer sufficient security to borrow the estimated £3,000 to £4,000 needed to finance the project nor was he able to lure building speculators to undertake the construction. In the end, Morton was forced to obtain a £2,000 loan from Lord Durham, which he was required to pay back within seven years of the date he began to receive rents.[45]

In the crucial area of company housing, therefore, viewers were both entrepreneurs and company agents. Yet there does not appear to have been a wide dichotomy of interests between viewers acting as housing speculators and those acting as employed, salaried managers. In both cases, the desire to maintain labor discipline could be served by depriving unruly workers of their homes. Perhaps because there was a general shortage of housing in the colliery districts during this period[46] or because there was the perception of a labor surplus after 1815, Morton never expressed any reluctance to turn out pit families for fear of reducing his rental income. Moreover, the lines between the responsibilities of owners as opposed to viewers for housing

---

(1980); Joseph Melling, "Employers, Industrial Welfare, and the Struggle for Work-place Control in British Industry, 1880–1920," in Howard F. Gospel and Craig R. Littler (eds.), *Managerial Strategies and Industrial Relations* (London: Heinemann, 1983).

43 Lambton Mss., Morton to Stephenson, 3 January 1835. See Michael Sill, "Landownership and the Landscape: A Study of the Evolution of the Colliery Landscape of Hetton-le-Hole, Co. Durham," *Durham County Local History Society*, 23 (1979), 2–11, who persuasively argues that local landowners had a decisive influence over the location of company housing in the second quarter of the nineteenth century. However, Sill does not address the question of who financed their construction.

44 N.C.L., Matthias Dunn's Diary, 26–7 January 1832.

45 Lambton Mss., Morton to Stephenson, 26 December 1835.

46 Flinn, *History of the British Coal Industry*, 429–30.

were obscured even further since the annual contract of employment for miners, the bond, included provisions regarding company housing. Still, the activities of viewers and agents in the provision of company housing in the coalfields serves to underscore their often ambiguous role as investors, entrepreneurs, and salaried employees.

The daily routine of a viewer was occupied to a great extent by the direct observation of the working of a colliery in order to evaluate the efficiency of production in terms of the utilization of labor, the employment of fixed capital, and the efficacy of on-site management. As Flinn noted, ideally a viewer was responsible directly to the owner.[47] He protected their investments by carefully planning the working out of a pit as well as its repair and maintenance. Matthias Dunn's diary, for example, is in good part concerned with the nature and extent of the workings of each pit in the Hetton Coal Company's collieries. The following portion of the entry for 10 November 1831 is typical:

> Down the Minor E. pit – Hutton Seam –. The E. workings are coming from two Pillar Districts.
> One – the So. Et. extremity of Shepherdson (about 15 score p. day) adjoining the Et. Boundary line.
> Other 15 score are coming from below the 8 feet Trouble (Mascall's Broken).
> These Boards all have been driven No. and So. – the Pillars left 22 yds. by 18 yds.
> These districts have been at work about 12 months and are getting away very clean – Bottom Coal 6 ins. thick sometimes curved and sometimes the contrary – but expected to be cast back.
> Allow the Hewers 1½ Corves p. score in lieu of small casting, altho' there is every reason to believe that very little is cast.
> This Pit is so very extensive that separate Overmen would appear quite advantageous.[48]

A viewer's responsibilities also entailed directing several ranks of supervisory personnel. At Hetton Colliery in 1831, there were seven overmen who oversaw the workings in the colliery's four working pits (P. E. H. Hair likens the overmen to shift foremen).[49] There were also three resident viewers and one underviewer at the pits as well as a colliery agent, two clerks, a bailiff, a surgeon, two heap inspectors, and several deputies.[50] Moreover,

---

[47] Ibid., 66.
[48] N.C.L., Dunn's Diary, 10 November 1831.
[49] P. E. H. Hair (ed.), *Coals on Rails: or The Reason of My Wrighting: The Autobiography of Anthony Errington from 1778 to around 1825* (Liverpool: Liverpool University Press, 1988).
[50] N.C.L., Dunn's Diary, n.d. (c. November 1831).

the colliery kept on staff a range of artisans needed to maintain the pits including an engine wright, smith, founder, sinker, and mason. At South Hetton Colliery in the early 1840s, there were twelve "superior officers" who supervised over five hundred employees both above and below ground.[51] A viewer's ultimate goal, nonetheless, was to secure the efficient production of coal, and in this he became the owner's representative in industrial relations with the miners.

To a great extent, as we shall see, workers possessed an extraordinary degree of autonomy at the point of production.[52] The relative autonomy of the worker necessitated the use of piece rates to secure both the requisite volume and quality of production. Of the four major grades of laborers in the mines – trappers, drivers, putters, and hewers – trappers and drivers were paid day wages, while hewers and putters were paid according to a complicated piece-rate system. Putting, or the manual hauling of coal from the face to the tramway, was remunerated based upon the number of corves transported over varying "renks," or distances to the tramway.[53] Hewers, the coal cutters, were paid a piece rate based upon the volume, not weight, of their production of coal. In the Northeast, piece rates were levied per score of corves, that is, the amount of coal equal to twenty 20-peck baskets in Northumberland, but sometimes twenty-one 20-peck baskets in Durham. (A "corf," or coal basket, was approximately twenty-six inches high and thirty-four inches in diameter.)[54]

Piece work in the mines was the subject of repeated bargaining and negotiations at the point of production because of the almost endless number of variations that might affect working conditions and hence earnings. Moreover, the establishment of relative piece rates was often dependent upon the disposition of different commodity markets thereby forging a link between the market and the labor process. Piece rates, for example, was different for working different seams of coal. At North Hetton Colliery, coal from the Main Seam in Moorsley West Pit was wrought at 6s. 6d. per score, while coal from the Moorsley East Pit working the Hutton Seam earned 5s. 8d. per score.[55] Higher prices per score were usually offered for the separation of large, round coal from small coal because the former was marketed at higher prices largely to the London consumer. Thus, at Cowpen Colliery,

[51] J. R. Leifchild, *Our Coal and Our Coal-Pits* (London, 1856; repr. New York: Kelley, 1968), 182–3.
[52] See Chapter 5.
[53] *Newcastle Chronicle*, 18 February 1832; N.E.I.M.M.E., Bell Collection, vol. 11, 402; Flinn, *History of the British Coal Industry*, 375–6.
[54] Flinn, *History of the British Coal Industry*, 461.
[55] N.C.L., Dunn's Diary, 16 November 1831.

Northumberland, separated coals were paid at the rate of 5s. per score, while unseparated coal was paid at only 3s. per score in 1830.[56]

Throughout the coalfield, hewers also received different piece rates based upon the distance they dug out in certain operations. Sometimes called "yard-work" or "narrow work," these payments were paid, as the term suggests, by the yard and applied most often to jobs such as winning new "headways" at the face or "holing" walls for new passageways. The yard "wand," or yardstick, which was used to measure this work became a symbol of the viewer's authority or tyranny.[57] Higher piece rates were also secured for working "in the Broken" (where the face already had been worked out and the miners dug out the remaining support pillars) and sometimes for working with Davy lamps (which were considered both dangerous and inconvenient).[58] At Cowpen Colliery, as in other collieries, higher rates applied as well to work in wet parts of the mine (but "only when Wet from the Roof") and for "double working," when two hewers worked side by side at the same bord (section of the face) instead of independently.[59]

Viewers and agents generally were free to bargain independently over piece rates with the pitmen. While it is true that the coal owners' cartel was revivified in 1805 partly in order to control rising piece rates and wages, these efforts, and others like them, repeatedly came to nought in the face of the immense variety of working conditions in different pits and collieries.[60] Even the appearance of a single, printed form for the bonds of all collieries in 1826 left both day rates and piece rates to be decided according to local circumstance.[61] The evidence that survives, particularly from periods of industrial actions and strikes, points to the commonplace nature of bargaining in the northern coalfield and the critical role of the viewer or agent in the bargaining process. The persistence of workplace bargaining in the northern coal industry throughout the first half of the nineteenth century reveals the secondary importance that the coal owners attached to the control of the production process. However, the coal owners' steadfast defense of the bond

---

[56] N.E.I.M.M.E, Watson Collection, Shelf 5/9/72.
[57] Dave Douglass, "Pit Talk in County Durham," in Raphael Samuel (ed.), *Miners, Quarrymen and Saltworkers* (London: Routledge & Kegan Paul, 1977), 348; Bill Williamson, *Class, Culture and Community* (London: Routledge & Kegan Paul, 1982), 95.
[58] D.C.R.O., Londonderry Papers, D/Lo/C 142(736), "To the Public."
[59] N.E.I.M.M.E., Watson Collection, Shelf 5/9/67.
[60] Flinn, *History of the British Coal Industry*, 355–6; P. E. H. Hair, "The Binding of the Pitmen of the North-East, 1800–1809," *Durham University Journal*, n.s., 27, no. 3 (1965).
[61] *A Candid Appeal to the Coal Owners and Viewers of the Collieries on the Tyne and Wear* (Newcastle-upon-Tyne, 1826), 9; see also the 1841 bond of Monkwearmouth Colliery reprinted in *Reports from Commissioners: Children's Employment (Mines)*, P.P. (1842), vol. 16, 536–8, in which the general printed forms omit piece rates and other particulars unique to individual collieries and pits.

and the Vend elucidates their determination to control the product and labor markets instead.

Bargaining and negotiations almost invariably took place between viewers or agents and the pitmen. Only in exceptional instances, such as was the case of Lord Londonderry in 1831 when he desperately needed cash, did owners intervene.[62] More commonly, such as in 1825 and at the beginning of Tommy Hepburn's union movement in March 1831, bargaining was initiated by the miners or their union when it recommended that the hewers at each colliery draw up a petition of grievances to present to their employers.[63] At Cowpen Colliery in 1831, for instance, two petitions were submitted to the agent, John Watson. Noticeably, these petitions were addressed "From the Workmen to the Agents of Cowpen Colliery."[64] That these petitions were considered valid forms of negotiation is witnessed by the careful manner in which Watson worked out his response. On one side of a page, a clerk listed the five demands of the miners; Watson's reactions and reasoning were listed on the other. Thus, the miners' demand for an advance of eightpence per score for double work was answered by "If the Boards [*sic*] are 4 yards wide to be paid, but Mr. W. does not apprehend there will be any during the ensuing year."[65] Similarly, the demand for a 4d. per score increase for working in the wet areas of the mine elicited the following comment: "Now paid, but only when Wet from the Roof."[66]

In another example characteristic of workplace bargaining, in June 1831, John Buddle sought to publicize the contractual problems with the miners at Wallsend Colliery.[67] He had printed and distributed a handbill that compared the demands of the men with the rates offered by the owners. While the handbill was designed to portray to the public a deadlocked dispute at Wallsend, Buddle's comments in the margin of his personal copy of the notice revealed a much more fluid situation. Publicly, he condemned the pitmen's demands for a variety of rate increases, but privately he continued to bargain. While openly and dogmatically rejecting the demand for a sixpence increase per score for separation, Buddle noted in the margin: "[The men] have agreed to take 4." Similarly, while publicly rejecting any change in the rate of fining for the amount of stone mixed in with the coal, privately

---

[62] See Chapter 7.
[63] The bulk of the famous pamphlet *A Voice from the Coal Mines; or, a Plain Statement of the Various Grievances of the Pitmen of the Tyne and Wear* (South Shields, 1825) is composed of a compilation of these petitions from different collieries; see also D.C.R.O., Londonderry Papers, D/Lo/C 142(658), Buddle to Londonderry, 6 March 1831.
[64] N.E.I.M.M.E., Watson Collection, Shelf 5/9/73.
[65] Ibid.
[66] Ibid.
[67] D.C.R.O., Londonderry Papers, D/Lo/C 142(736).

he wrote, the men "have agreed to pay 3d. for 3 quarts and 6d. for 4 quarts." Further demands by the miners, Buddle noted, were given up during the course of negotiations.

The social and economic basis of these demands is often impenetrable. In some cases, miners' demands certainly reflected the attempt to recapture lost privileges. For example, the demands of Cowpen's workmen for the reinstitution of candle money in 1831 – a twenty-six shilling per annum allowance to purchase candles that had ended at that colliery two years previously – is one such event.[68] Other demands common in the 1820s and 1830s were affected by the rise of a cadre of union leaders who exerted a special influence over the pitmen's bargaining demands.[69] Thus, the demands to reduce the hours of children's work and to separate housing from the terms of employment seem to have filtered down from the union leadership rather than up from the rank and file. Still other demands reflected changes in the nature of work. The demand for higher piece rates for working with Davy lamps was evidence of the greater difficulty and danger experienced from working with them.

However, there seems to be no clear reason why working with "Davys in the Broken" merited a sixpence per score increase in the early 1830s whereas the difficult job of winning a new section of the face – Dave Douglass still calls this "hideous work,"[70] – was subject to only a threepence per yard increase. Or, for example, why working in wet sections of a mine required in some collieries a fourpence per score increase rather than three- or six-pence.[71] There indeed seems to be little normative reason for the exact rates chosen, a point Douglass reiterates,[72] except that they were part of a deep matrix of bargaining between miners and viewers.

One consequence of management's relative inability to control the point of production is evidenced by the viewer's frequent recourse to the work experience of miners and other underground laborers in matters of safety. It is apparent from Anthony Errington's autobiography that John Buddle frequently sought out the experience and expertise of underground workers when faced with questions concerning the safety of working conditions.[73] In one instance, a safety "society" was established at Percy Main Colliery composed principally of overmen, deputies, wastemen, and hewers to discuss

---

[68] N.E.I.M.M.E., Watson Collection, Shelf 5/9/67.

[69] See Chapter 7.

[70] Douglass, "Pit Talk in County Durham," 347.

[71] N.E.I.M.M.E., Watson Collection, Shelf 5/9/73.

[72] Dave Douglass, "The Durham Pitmen," in Samuel (ed.), *Miners, Quarrymen and Saltworkers*, 239.

[73] Hair, *Coals on Rails*, 86–90.

remedies to the "dangrus State" of the pits.[74] At Hetton Colliery in the early 1830s, the hewers were first appraised of a safety report on the worked-out portions of the mine and subsequently approached the managing committee of the colliery to express their concerns.[75] (In the latter case, it is even likely that the report was prepared by one of the miners' delegates, Ralph Dove, although this cannot be proved.)[76]

The negotiated workplace relationship between management and labor established through the practice of piece-rate bargaining was limited principally by the imposition of an array of fines. Fines acted largely as a substitute for direct control at the point of production. Thomas Crawford, viewer of the earl of Durham's collieries in the mid-1840s, explained: "Fines are absolutely necessary; without them we could not cause the regulations [of the colliery] . . . to be strictly observed."[77] Similarly, T. J. Taylor, a coal owner and nephew of the chairman of the Coal Trade Committee, argued:

> In levying them [fines] the object of the coal-owner is to prepare the article produced in as good a state as possible for the market; and they do, therefore, not only affect him as competing with his neighbours, but the public at large, as being most essentially concerned in the production of good coal, *i.e.*, of coal which is not mixed, "fouled," with stones, slates, or small coal.[78]

The most common fines, therefore, were imposed for "Foulness," that is, the inclusion of stones or slate in the corf. John Watson, viewer of Cowpen Colliery, was of the opinion that a fine of between sixpence and one shilling was common for foul corves.[79] Fines also were levied for sending corves to

[74] Ibid., 89.

[75] N.C.L., Matthias Dunn's Diary, 29 November 1831; 1 December 1831.

[76] The evidence for this is internal in nature. "Ra. Dove" certainly carried out the inspection of the Hetton Colliery wastes along with another man named Charlton. Dunn's account of the pitmen's reaction to the report notes that Charles Parkinson, one of the most well known union leaders under Tommy Hepburn, led a delegation of eight workers to meet with management. When the delegation's protests were not accepted, Dunn noted: "Parkinson has about lost their [the men's] confidence as well as Ra. Dove." This may indicate that Dove, like Parkinson, was a union delegate. Dunn's list of colliery officers compiled about two weeks earlier makes it clear that he was not an overman nor was he employed in any of the upper ranks of colliery administration. N.C.L., Matthias Dunn's Diary, 10 November 1831; 1 December 1831.

[77] *Report of the Commissioner appointed under the Provisions of the Act 5 & 6 Vict., c. 99, to Inquire into the Operation of that Act and into the State of the Population in the Mining Districts, 1846* (hereafter, *Report of the Commissioner . . . into the State of the Population in the Mining Districts, 1846*), P.P., 24 (1846), 63.

[78] *Report of the Commissioner . . . into the State of the Population of the Mining Districts, 1846*, P.P., 24 (1846), 11.

[79] N.E.I.M.M.E., Watson Collection, Shelf 5/9/76c.

the surface that were not filled properly. (Needless to say, a common complaint among miners was that they should not be responsible for the settling
or spillage of coal after it left their control.)[80] Finally, there was a series of
fines for poor work practice, such as the improper use of Davy lamps or
improper techniques used at the face.[81]

Despite the common claims that fines were imposed "despotically" and
"tyrannically," terms used repeatedly in the famous pamphlet *A Voice from
the Coal Mines* (1825), it was the distinct separation of ownership from
management that helped mask the "structured antagonism" of capital and
labor.[82] It was commonly argued by union publicists that management severed the natural identity of interests between capital and labor by misrep-
resenting their demands and by exercising an unwarranted measure of
authority. In the midst of the 1831–2 strike, a union advocate complained
in the local newspapers that a satisfactory settlement was being hindered by
the owners' separation from the direct management of their concerns:

> One of the greatest evils which had befallen the men in all the controversies
> which they have had with their owners has been that their owners viewed
> everything through the medium of their agents. Had the owners themselves
> come more frequently in contact with the men, they would have better under-
> stood their real situation, and they would have learnt to put the right value
> upon the evidence of interested viewers and agents.[83]

An open letter to the northern coal owners published during the 1844
miners' strike further reveals the powerful impact made by the separation
of ownership from management. William Mitchell wrote:

> Rebellion is the wages of tyranny and I am sorry to have to state, that the
> conduct of your servants, but our masters, the viewers, has, with some hon-
> ourable exceptions, been as mean and despotic as that of the petty princes of
> Germany. Having held a delegated, but yet an almost absolute despotic power
> they have, too generally, governed us in an arbitrary and tyrannical manner.

By 1844, however, the miners were not as sanguine as they once had been
about the role of ownership in labor relations. Mitchell aptly reasoned that
the owners were "like the absentee landlords of Ireland, [who] know little

---

[80] *A Voice from the Coal Mines*, 27.
[81] Flinn, *History of the British Coal Industry*, 378–9.
[82] Edwards, *Conflict at Work*, 5.
[83] *Newcastle Chronicle*, 24 March 1832.

of, and that little knowledge tends to produce little regard for, those from whose labour you derive your incomes."[84]

The ambiguity of the sources of authority in a multilevel firm was reflected in the language of these struggles as well. The term "Masters" often applied equally to viewers and agents as well as to the legal proprietors of the collieries. Robert Atkinson, a union leader of the 1830s, proclaimed that one of the most beneficial results of unionization was the fact that miners "were now no longer in awe of the humours of their masters." By "masters," however, he did not mean the owners of capitalist enterprise but the salaried viewers: "The time was," Atkinson went on to say, "when the viewers came down the pits, the questions were, 'What sort of humour is he in? can he be spoken to today, think you?' But now they could tell them manfully of their grievances, where there were any, and request their redress."[85] Tommy Hepburn, the leader of the 1831–2 miners' union, explained that the unique power of the viewers rested upon their control of both employment and housing. "Thus," he said, "are we made slaves to the tempers and caprices of the Viewers – and the workmen are fearful of even consulting with each other on their grievance lest it might come to the ears of their masters, and they be deprived of both bread and shelter."[86]

In one instance, the position of the viewer himself became the focus of industrial dispute. At Hetton Colliery, the seat of the union movement of the 1830s, Matthias Dunn attempted to institute a system of separation to improve the market value of coal. In return for separating small from round coal, Dunn offered to increase piece-rate prices from 5s. to 6s. per score in the Hutton Seam and 6s. 3d. to 7s. per score in the Main Coal Seam. On 16 November 1831, he duly met with four delegates from the pitmen of the colliery to negotiate the changes. His diary reports that "a great deal of Discussion ensued but a meeting [with the men] is necessary before any decision can be come to upon the subject."[87]

The miners at Hetton rejected the proposal in large part because the fines for poor separation were exceedingly high. By December, Dunn's frustration began to become apparent, and he toyed with the idea of entirely stopping work at the Elemore Pit. In so doing, he could transfer the twenty hewers there to other pits "and by filling the other pits beyond what they are now in a great measure *enforce* the stowing of Small Coal which cannot be done

---

[84] William Mitchell, *The Question Answered: "What do the Pitmen Want?"* (Bishopwearmouth, 1844; repr. New York: Arno, 1972), 21.

[85] *Newcastle Chronicle*, 23 June 1832.

[86] N.E.I.M.M.E., Bell Collection, vol. 11, 218, "An Account of the Great Meeting of the Pitmen of the Tyne and Wear on Newcastle Town Moor, on Monday, March 21, 1831."

[87] N.C.L., Dunn's Diary, 16 November 1831.

satisfactorily as at present."[88] However, on December 15, the owners' committee of the Hetton Coal Company rejected Dunn's idea because "the men might object to have such an arrangement carried into Effect, and stick [strike] that therefore it had better be deferred."[89]

When the renewal of the annual bonds commenced in March 1832, the dispute still rankled both workers and management. Dunn described the miners as in "great bad humour," and they "cavilled with almost every clause [of the bond]."[90] On 17 March, a deputation of the pitmen sought to enter a meeting of the owners' committee of the company. After being refused entry, the deputation submitted to them a petition demanding the discharge of Matthias Dunn.[91] While the owners' committee rejected the petition "with suitable disgust," according to Dunn, the actions of the miners were considered so extraordinary that the matter was referred to the executive committee of the cartel, the General Meeting of the Coal Owners of the Tyne and Wear. Dunn assured the meeting that the dispute was simply one between masters and men, and the meeting resolved:

> that the statement made by Mr. M. Dunn respecting the claims by the Hetton Men is perfectly satisfactory; and that the approbation by this Meeting of Mr. Dunn[']s conduct, be conveyed to the Owners of that Colliery, with a request that they will resist every attempt by their workmen, to interfere with the appointment of those Gentlemen to whom they may think proper to confide the management of their concerns.[92]

The district union leadership, under Tommy Hepburn, fully accepted both the decision and the rationale of the committee of coal owners. Hepburn agreed that the men's protest may have been warranted, but their actions were nonetheless "unreasonable" and constituted "an interference with the rights of the Owners."[93] Hepburn's official position was that the "demand was made by a few indiscreet individuals, contrary to the wish, and without the consent of, the body at large."[94] As Dunn's case illustrates, the managerial functions of the viewers helped shield ownership from direct confrontation with labor. Rather than question the social basis of ownership and production, as may happen when owners are directly confronted by labor, management itself became the subject of negotiation.

---

[88] Ibid., 14 December 1831.
[89] Ibid., 15 December 1831.
[90] Ibid., 17 March 1832.
[91] Ibid., 19 March 1832.
[92] Ibid., 24 March 1831; N.R.O., General Meetings of the Coal Owners of the Rivers Tyne & Wear, 24 March 1832, 123–4; *Newcastle Chronicle*, 5 May 1832.
[93] N.E.I.M.M.E., Bell Collection, vol. 11, 416, "An Address to the Public."
[94] Ibid.

Given their critical role in both the organization of production and industrial relations, it is not surprising that viewers and agents often became the subject of direct popular action during strikes. John Buddle was hung in effigy at the beginning of Tommy Hepburn's strike.[95] The viewer's house at Cowpen Colliery was broken into during the 1831 strike, and his cellar was raided.[96] Intimidation of the viewers of collieries certainly was a regular feature of the strike led by Tommy Hepburn in the 1830s. John Buddle complained to Lord Londonderry that "a system of Terror and annoyance has been adopted against the Viewers and Agents of all Collieries which have not complied with the demands of the men." These popular actions commonly took the form of a charivari, Buddle continued, where "numbers of men assemble in the night dressed in Women's Clothes, fire Guns and pistols, break their Windows, destroy their Gardens, etc., etc."[97] Wherever the viewers went during the course of the strike they were hounded by the pitmen. It was Buddle's experience that the pitmen "beset the Roads in all directions, and insult all Colliery Agents wherever they go.... I am beset, hooted and hissed and my life threatened wherever I go, not only by the Pitmen, but by all the rabble of every description."[98]

The stabilizing effect of the separation of management from ownership in the coalfields is perhaps epitomized, however, by the case of Lord Durham. The Lambton family was perhaps the largest proprietor in the northern coalfield throughout the nineteenth century. The active participation of J. G. Lambton in the formulation of the First Reform Bill earned him the sobriquet "Radical Jack." However, the administration of Lambton's collieries was hardly radical. His agent, Henry Morton, was a strict taskmaster as well as a pedestrian advocate of Malthusianism.[99] Morton's first response to the formation of the Hepburn's union in 1831 was typical of the mental world he inhabited: "There are considerably *more* men, both upon the Tyne and Wear than can be fully employed – therefore it is quite impossible they can succeed in their demands."[100]

Morton's acute lack of vision in this case was matched only by his dedication to breaking the union once it became established. Morton was the first agent to reopen a pit after the strike commenced in April 1831, when

[95] D.C.R.O., Londonderry Papers, D/Lo/C 142(682), Buddle to Londonderry, 16 April 1831.
[96] Richard Fynes, *The Miners of Northumberland and Durham* (Blyth, 1873; repr. S. R. Publishers, 1971) 21; the threatening letter that was left behind is also quoted in Thompson, *Making of the English Working Class*, 715.
[97] D.C.R.O., Londonderry Papers, D/Lo/C 142(714), Buddle to Londonderry, 28 May 1831.
[98] Ibid., D/Lo/C 142(711), Buddle to Londonderry, 19 May 1831.
[99] Morton was "the sworn foe of the slovenly farmer and the striking pitman" according to Spring, "Agents to the Earls of Durham," 107.
[100] Lambton Mss., Morton to Durham, 23 March 1831.

Lumley pit began work under a guard of troops on April 27.[101] After reaching a settlement with the striking miners in May, Morton continued to harass and victimize the local union leaders. He threatened to evict union deputies from their homes but failed only when the local magistrates refused to execute the warrants for eviction.[102] Throughout the summer of 1831, the Lambton collieries continued to be the scene of unrest; miners from both Newbottle Colliery and Lambton Colliery were prosecuted for riot and assault in June of that year.[103] Relations between the miners and Morton deteriorated to such an extent that the men dug a mock grave for him at the height of the strike.[104] Yet in spite of the accumulated tensions between Morton and the pitmen, this ill will did not affect Lord Durham's reputation. On the contrary, at the great victory rally of Hepburn's union on Boldon Fell in August 1831, Robert Arkle, a union leader, publicly labeled Lord Durham, without irony, as "the pitmen's friend."[105]

The other great Durham landowner cum colliery entrepreneur, Lord Londonderry, at times also worked hard to maintain the ties of "good lordship" across the gulf that separated his seat at Wynyard from the mines. In 1845, after the defeat of the Chartist National Miners' Association, Londonderry directed Buddle's successor, Neville Hindhaugh, to extend "measures of Comfort and advantage to [the pitmen] as will incontestably prove that we, on our part, desire only that reciprocal good understanding and mutual Benefit that should ever exist between the Employer and the Employed." In so doing, Londonderry hoped to "rely firmly on its cementing those old Ties of Attachment that have so long prevailed in the Old Collieries of the Vane and Tempest families between the Pitmen and their Masters."[106] However, the countess of Durham's colliery doctor, William Morrison, implied that these measures of paternalism were likely to have little effect because owners did not manage their concerns, they were not resident in the colliery district, and most viewers had little interest in the improvement of the "defects in the habits and characters of pitmen:"

> The owners of collieries know themselves, little of the detail of the management of their concerns. They do not reside much in the vicinity of their works, nor take any part in their management.... The larger collieries are entirely man-

[101] Ibid., 27 April 1831.
[102] Ibid., 19 May 1831.
[103] Ibid., 29 June 1831.
[104] Spring, "Agents to the Earls of Durham," 110.
[105] Newcastle Chronicle, 20 August 1831.
[106] D.C.R.O., Londonderry Papers, D/Lo/C 739(2), Londonderry to Hindhaugh, 17 June 1845.

aged by viewers, who, it must be confessed, do not take that interest...which might be expected from them.[107]

To a large extent, Morrison's analysis was correct, but his conclusions were quite wrong. The separation of ownership from management did not ultimately lead to social calamity. The bifurcation of industrial authority actually exculpated the aristocracy and large capitalists from responsibility for industrial relations and allowed them to act as disinterested arbitrators in disputes between viewers and miners. Therefore, the separation of ownership from management was a force acting toward social stability, not instability.

The viewers, however, came to play a crucial role in the development of social relations in the northern coalfield principally because they not only were the source of authority in matters of work and housing, but they also served to mediate the acceptance of bargaining as the principal fulcrum of labor relations. The ultimate significance of the role of the viewers and agents to the miners was testified to by a unique adaptation of high political culture. Upon the successful completion of the 1831 strike, the miners at Wallsend Colliery chaired their viewer, John Buddle, in the manner of a successful parliamentary candidate:

> I have been demeaned and hunted like a wild beast by the infuriated body of Pitmen, while they were in a State of excitement. But now, that they are returning to something like a State of tranquility, they are doing me justice by every sort of apology and acknowledgment of the injustice they have done my character during this *phrensy* [*sic*]. I am complimented for having "fought them fairly, *like a Man*." I was chaired *by force* last Friday evening at W[alls]end and was obliged to *abscond* on Sat. morning to avoid the *honor* of being drawn in grand procession from W[alls]end to Newcastle. So much for popular odium, or popular favour.[108]

The actions of the pitmen of Wallsend reveal not only the social significance of the viewer, but also the degree to which bargaining was an accepted part of the landscape of northern industrial relations. A fair, manly fight epitomized the ethic of early industrial pit culture. The prerogatives of capital could be safely defended from behind the walls of management; and, in fact, they could be celebrated by labor when market forces were deemed

---

[107] *Reports from Commissioners: Children's Employment (Mines)*, P.P. (1842), vol. 16, 730.
[108] D.C.R.O., Londonderry Papers, D/Lo/C 142(739–40), Buddle to Londonderry, 23 June 1831. Similarly, Buddle's nephew, the viewer at Percy Main Colliery, was forced to "go with the Pitmen to *carouse* with them, in celebrating their Victory." D/Lo/C 142(761), Buddle to Londonderry, 14 July 1831. And at Fawdon Colliery, the colliery agents were chaired by the pitmen's wives. *Newcastle Chronicle*, 18 June 1831.

to have operated fairly and equitably. There was, needless to say, a prolonged struggle along a frontier of control that made up the substance of labor history during this era. However, as we shall see, this was less a struggle against the "real" subordination of labor at the point of production than a struggle to control the terms of exchange relations in the marketplace.

4

Family, community, and the labor market

The labor market of the nineteenth-century coal industry profoundly influenced family life and community. That work should influence life is not in itself a striking observation, especially in a colliery village with company-owned housing. However, that the market for labor structured the region's demographic profile, community organizations, and gender relations points to the extent to which work and life were intertwined outside of the formal workplace.

Management's inability and unwillingness to control the point of production shaped the way in which work influenced life. Yielding substantive influence over the labor process, management and ownership instead relied upon their ability to influence the terms of exchange within the labor market. In part, this function was performed by their control of the system of annual contracts, including housing, as well as by the collusion of coal owners to restrict mobility. However, these coercive efforts of capital were matched by complementary tactics of accommodation. Particularly after the industrial unrest of the 1820s, there was a marked revival of interest in the functions of industrial paternalism. This new concern with education, housing, and religion should not be construed solely as a reflection of the desire to inculcate deferential attitudes among workers and thereby achieve ideological domination.[1] Instead, the emergence of industrial paternalism signaled the transference of the terrain of struggle away from the point of the production and into the market. The paternalism of the second quarter of the nineteenth century, therefore, was part of an effort to attach workers to the firm, limit labor mobility, and establish control of the labor market.

The social impact of the labor market

The claim that a labor market existed must be both qualified and defended. Mining was a most insular trade, and both employers and workers generally

[1] Patrick Joyce, *Work, Society and Politics: The Culture of the Factory in Later Victorian England* (Brighton: Harvester, 1980); H. I. Dutton and J. E. King, "The Limits of Paternalism: The Cotton Tyrants of North Lancashire, 1836–54," *Social History*, 7, no. 1 (1982), 59–74.

accepted the premise that the best workers were "natural bred pitmen."[2] The Poor Law Commissioner for Durham and Northumberland, John Wilson, wrote in his report of 1834 that "pitmen must be bred to their work from childhood; they cannot well be drilled to it at a later age; their numbers cannot be recruited from any other class."[3] This observation is clear evidence of the existence of a segmented labor market based on birth, gender, training, and experience.[4] Yet given the employers' preference for locally trained male labor, the expansion of coal production in the nineteenth century necessarily placed a premium upon experienced labor and, therefore, allowed labor to operate along market lines, despite significant impediments.[5] In fact, as will be argued in later chapters, the principal goal of early-nineteenth-century trade unions among miners in the Northeast was to secure equal power in the labor market by removing market impediments that benefited the employing class.

Many labor market impediments were daunting nonetheless. Before the institution of a monthly contract in 1844, miners were employed on an annual basis according to the terms of the bond.[6] Thus, a labor market tended to operate for only about a fortnight each year during "binding time." Even during binding time, there were critical social forces that militated against the operation of a competitive labor market. Perhaps the most significant of these forces was that of housing. The termination of a miner's bond meant that he and his family yielded their claim to a company house; the end of a job entailed the relocation or eviction of the family. Both management and miners were acutely aware of the dual power exercised by the viewers through the provision of housing and employment and, more importantly, how company-tied housing constrained their ability to bargain with capital. As early as the 1830s, in contrast to Colls's view, housing was a crucial concern of union movements.[7] Hepburn's union, for example, demanded the separation of housing from the terms of employment in 1831, and the owners subsequently agreed to give miners a fortnight to leave their homes after the end of the bond.[8] When this extra fortnight's respite failed

---

[2] Flinn, *History of the British Coal Industry*, 339–49.
[3] *Reports from Commissioners: Poor Laws*, P.P., vol. 28 (1834), appendix A, no. 5, 130a.
[4] D. M. Gordon, R. Edwards, and M. Reich, *Segmented Work, Divided Workers* (Cambridge University Press, 1982); Glen G. Cain, "The Challenge of Segmented Labor Market Theories to Orthodox Theory: A Survey," *Journal of Economic Literature*, 14 (1979), 1215–57; William Reddy, *The Rise of Market Culture: The Textile Trade and French Society 1750–1900*, (Cambridge University Press, 1984); Joyce, *Historical Meanings of Work*.
[5] Flinn, *History of the British Coal Industry*, 340–1.
[6] Generally, see ibid., 352–8, and Colls, *Pitmen of the Northern Coalfield*, 45–51.
[7] Colls, *Pitmen of the Northern Coalfield*, 262–3, argues that "the housing problem was never central to trade union activities until late [nineteenth] century."
[8] N.E.I.M.M.E., Bell Collection, vol. 11, 217.

to protect labor from intimidation and victimization in 1832,the union sought legal advice, perhaps from Robert Lowery, the radical and later Chartist from North Shields, as to the rights of miners to their houses upon the expiration of the bond.[9]

The operation of a labor market was similarly inhibited by the active collusion of coal owners. The labor shortages in the Northeast engendered by the Napoleonic Wars brought to fruition an active and competitive labor market. Hewers could demand and receive a bonus, called "binding money," of up to twenty guineas by 1805.[10] In that year, Flinn estimated, labor turnover reached a peak of around 30 percent.[11] However, in September 1805, the coal owners met to limit competition among themselves for labor and to reduce the level of binding moneys.[12]

Binding money was immediately reduced, and the market forces for labor were reined in, but the dialectical tendency of coal owners to both collude and compete could not be eradicated so easily. Subversion of labor market impediments was often in the best interest of some coal owners. The experience of many collieries throughout the period of rapid expansion after the early 1820s led not only to the intermittent revival of binding money, but also to the active poaching of labor, revivifying the market for labor. For example, at the conclusion of the miners' strike in June 1831, six collieries that could not come to an agreement with the men demanded an indemnity to be drawn from the funds of the coal owners' cartel.[13] When the cartel hesitated, John Buddle reported that the representatives of the six collieries "came in self-defence to a determination, if they could do no better, to offer a Bounty of 10 guineas per Man for Hewers."[14] With no support from the cartel forthcoming, at least one of the six collieries, Callerton, signed on workers by offering two guineas per man in binding money.[15] Perceiving the threat to their control of the labor market, in August the cartel agreed to indemnify collieries that were still being struck by their workers.[16]

The poaching of labor, or as Flinn put it, the "strong temptation to look covetously at his neighbour's work-force,"[17] was still another way in which

[9] *Newcastle Chronicle*, 23 June 1832.
[10] Flinn, *History of the British Coal Industry*, 355–6.
[11] Ibid., 343–4.
[12] Ibid., 356.
[13] N.R.O. General Meeting of the Coalowners of the Rivers Tyne and Wear, 13 June 1831, 109.
[14] D.C.R.O., Londonderry Papers, D/Lo/C 142(733), Buddle to Londonderry, 13 June 1831.
[15] Ibid., D/Lo/C 142(735), Buddle to Londonderry, 18 June 1831.
[16] N.R.O., General Meeting of the Coalowners of the Rivers Tyne and Wear, 5 August 1831, 110.
[17] Flinn, *History of the British Coal Industry*, 340.

the coal owners' control of the labor market was undermined. Although poaching was outlawed according to the terms of the 1805 coal owners' agreement, it remained widespread nonetheless. The Hetton Coal Company, for example, which expanded rapidly in the decade after 1825, was notorious for poaching both child and adult labor. The bitter feud between the Hetton Company and J. G. Lambton was inflamed by the former's massive poaching of miners.[18] In 1825, Buddle wrote to Lord Londonderry that "the *Hettonians* have made almost a clean sweep of the best of Lambton's families of pitmen from Newbottle, and nearly deprived him of the *putter* Boys; they have also carried off a great many from us. Report says they have bound 100 Hewers more than their former complement."[19]

Certainly the strongest evidence for the existence of a regional labor market is the extraordinarily high degree of local mobility that was exhibited by miners and their families. At Wallsend on Tyneside between 1798 and 1812, for instance, only 8 percent of the 667 male residents for whom information exists were actually born in the parish. Over two-thirds of the men had migrated to Wallsend from within a radius of ten miles. Of the female residents, less than 15 percent of a total of 688 women had been born in Wallsend.[20] Sill's work on Hetton parish reveals a similar pattern. Of the 641 coal miners who were heads of households in 1851, only 3.59 percent had been born in the parish. Over 81 percent of the miners had been born elsewhere in the northern coalfield, and most of those had migrated from other collieries that were within several miles of Hetton.[21]

Furthermore, government evidence from the first half of the nineteenth century supports the contention of a mobile and active labor force. The famous parliamentary investigation of children's employment in mines in the 1840s revealed a pattern of mobility influenced predominantly by the difficulty of work and piece-rate prices, but also by the provision of housing, gardens, and children's employment. George Johnson, a viewer of three Tyneside collieries, explained that "the first thing that induces pitmen to move is the nature of the work, as influencing prices; and then the nature of the houses and gardens."[22] One of the miners interviewed by the commissioners, forty-

---

[18] See above, Chapter 2.

[19] D.C.R.O., Londonderry Papers, D/Lo/C 142(147), Buddle to Londonderry, 23 March 1825.

[20] This information was generously provided by Rab Houston and Trillia Scoins of the Cambridge Group.

[21] Michael Sill, "Mid-Nineteenth Century Labour Mobility: The Case of the Coal-Miners of Hetton-le-Hole, Co. Durham," *Local Population Studies*, 22 (1979), 44–50; see also Flinn, *History of the British Coal Industry*, 341–7.

[22] *Reports from Commissioners: Children's Employment (Mines)*, P.P. (1842), vol. 16, 569.

two-year-old Luke Gray, also claimed that "men have to leave the colliery sometimes because their boys are put too young to this work."[23]

Miners' biographies and autobiographies, even though they tend to be of labor activists, provide further evidence of the mobility of the labor force and the concomitant existence of an active labor market. John Wilson, the founder of the Durham Miners' Association, had shifted collieries six times before he was twenty-three years old.[24] Thomas Burt, who became one of the first working-class M.P.s, said that his father had "the wandering instinct," and he had moved their family at least ten times.[25] E. A. Rymer worked at more than a dozen pits in several different counties during the course of his mining career.[26]

While Thomas Burt may have believed that his father randomly wandered the coalfield to fulfil a vague instinct, there were nonetheless broad patterns of population movements that reflected labor markets. Arthur Smailes likened population changes in the northern coalfield to a life cycle: An initial spurt of population growth led by immigration and economic activity marked a colliery town's youth. Subsequently, the town matured, its population became sufficient to meet the labor needs of the colliery, and immigration declined. When the mines began to be exhausted, a period of depopulation set in as miners left in search of better work; rejuvenation might occur (here the life-cycle metaphor ends) if new collieries were established or new seams of coal tapped. Otherwise, the town experienced a period of decline and then death.[27]

However, one can augment Smailes's observations by realizing that these population movements were defined by specific gender relations and that the four stages of development were also characterized by specific types of social relations. During the initial phase of expansion, population growth was dominated by young, adult males. The town of Haswell, for example, grew from a population of 114 in 1801 to 263 in 1831.[28] When the Haswell

---

[23] Ibid., 584.

[24] John Wilson, *Memories of a Labour Leader: The Autobiography of John Wilson, J.P, M.P.* (London: Unwin, 1910).

[25] Aaron Watson, *A Great Labour Leader: Being the Life of the Right Honourable Thomas Burt* (London: Brown, Langham, 1908).

[26] E. A. Rymer, "The Martyrdom of the Mine," reprinted in *History Workshop Journal*, 1 and 2 (Spring and Autumn, 1976).

[27] Arthur E. Smailes, "Population Changes in the Colliery Districts of Northumberland and Durham," *Geographical Journal*, 91 (1938), 220–32.

[28] *Abstract of Answers and Returns, made pursuant to an Act of 2 Geo. 4, for taking an Account of the Population of Great Britain, and of the Increase and Diminution thereof* (hereafter, *1831 Census*): "Comparative Account of the Population of Great Britain in 1801, 1811, 1821, 1831," P.P. (1833), vol. 36, 87.

Coal Company and the South Hetton Coal Company opened collieries in the area in 1833 and 1835, local population soared.[29] By 1841, Haswell's population jumped to 3,981 – an increase of over 1,400 percent from 1831[30] – with about 14 percent more men than women.[31] Similar effects on gender relations can be seen in other boomtowns of the industrial era. During the 1830s, several collieries opened in Kelloe parish, a sparsely inhabited area of east Durham. The population of the parish's town of Coxhoe jumped from 154 in 1831 to 3,904 a decade later. The neighboring town of Thornley grew from only 50 inhabitants to 2,730; nearby Wingate from 115 people in 1831 to 2,625 in 1841.[32] In all three towns, men outnumbered women by significant margins: in Coxhoe by 15.2 percent, in Thornley by 12.1 percent, and in Wingate by 20.4 percent.[33]

Social and industrial relations in these new boomtowns were generally less stable and more contentious than elsewhere in the coalfield. Thus, it is not surprising that Hetton, the largest and fastest growing colliery of the late 1820s and early 1830s, was the center of Tommy Hepburn's union movement or that Thornley Colliery was the colliery hardest hit by the 1844 miners' strike. Contemporaries certainly believed that new mining towns attracted a "less respectable" class of workers. Seymour Tremenheere, the mines commissioner, explained the tenacity of the strikers at Thornley Colliery with reference to "its being a comparatively new work, and consequently resorted to by the more unsteady characters from other neighbourhoods."[34] William Hunter, the viewer of Walbottle Colliery, put the phenomenon more bluntly when referring to Sacriston and Charlaw collieries near Chester-le-Street: "It is a new colliery and the scum of the others was there."[35] Indeed, historians must be wary of too broadly employing anachronistic concepts of community in order to explain the social foundations of early industrial protest.[36] In Durham, regional strikes generally did not originate from communities in which social, familial, and institutional ties were either strongest or of long standing. Instead, the foci of protest were from precisely those

---

[29] Galloway, *Annals of Coal Mining*, 451.

[30] *Abstract of Answers and Returns under the Population Act, 3 & 4 Vic., c.99* (hereafter *1841 Census: Enumeration Abstract*), part I, P.P. (1843), vol 22, 85.

[31] *1841 Census: Enumeration Abstract*, part I, P.P. (1843), vol. 22, 85.

[32] *1831 Census: Comparative Account*, P.P. (1833), vol. 36, 88; *1841 Census: Enumeration Abstract*, part I, P.P. (1843), vol. 22, 85.

[33] *1841 Census: Enumeration Abstract*, part I, P.P. (1843), vol. 22, 85.

[34] *Report of the Commissioner... into the State of the Population of the Mining Districts, 1846*, P.P. (1846), vol. 24, 22.

[35] *Reports from Commissioners: Children's Employment (Mines)*, P.P. (1842), vol. 16, 617.

[36] Craig Calhoun, *The Question of Class Struggle: Social Foundations of Popular Radicalism During the Industrial Revolution* (Chicago: University of Chicago Press, 1982).

communities where such ties were weakest; communities, in a sense, more akin to Marx's rootless, urban masses than the populist artisans of village radicalism.

A colliery village's period of population consolidation, what Smailes called "maturity," brought a more equal distribution of men and women along with a more stable community. Ralph Elliot, underviewer of the Londonderry collieries, claimed that employment at the older mines of Pensher and Rainton was particularly steady. "At Pensher," he said, "there are not 10 men who have not belonged to the colliery all their lives. At Rainton, two-thirds of the number have also been with us all their lives; the rest are new hands, the works there having been increased of late."[37] Correspondingly, the total population of West Rainton, the location of Rainton Colliery, was remarkably stable. The revitalization of that colliery was one of John Buddle's more important projects after Londonderry's marriage to Frances Anne.[38] Between 1811 and 1821, the town nearly doubled in size from 629 inhabitants to 1,160. However, through the next two decades the population of West Rainton barely changed: In 1831, the town had a population of 1,184 and ten years later, 1,054.[39] The number of men and women was nearly equal by 1841, indicating a more stable community life; the census counted 542 men and 512 women, a difference of less than 6 percent.[40]

While the period of a town's maturation brought the number of men and women into greater balance, the gradual decline of a colliery town witnessed emigration led by men and, often as a result, the balance of men and women shifted in favor of the latter. Mining communities at Pensher, one of Lord Londonderry's "steady" collieries, and Houghton-le-Spring were the homes of more women than men by 1841. Not surprisingly, this gap between males and females was narrow during childhood and old age but widened considerably during the adult years when male workers were most liable to emigrate. For example, in 1841 Houghton-le-Spring had nearly 7 percent more women than men; Pensher just over 2 percent more women than men. In age groups over twenty, however, the gap widened significantly, clearly indicating male-dominated emigration. In Houghton-le-Spring, there were only 1.6 percent more women under twenty years of age, but over 12 percent more women than men over twenty. At Pensher, there were actually 7 percent more men than women under twenty (this might be accounted for by the fact that older

[37] *Report of the Commissioner... into the State of the Population of the Mining Districts, 1846*, P.P. (1846), vol. 24, 15.
[38] Sturgess, *Aristocrat in Business*, 30–7.
[39] *1841 Census: Enumeration Abstract*, part 1, P.P. (1843), vol. 22, 85.
[40] Ibid.

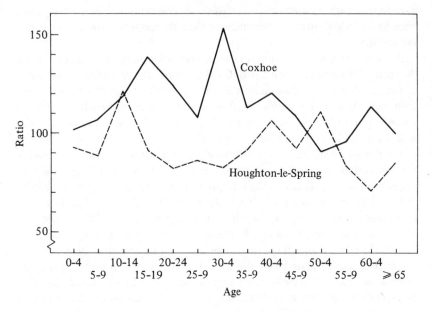

Figure 4.1. Age-specific sex ratios, 1841. (From *1841 Census: Age Abstract*, P.P. [1843], vol. 23, 74–5.)

mines with longer distances from the face to the shaft needed more haulers, or putters, who were almost invariably adolescent males), but more than 11.5 percent women than men over twenty.[41]

Different age groups thus differed markedly in their ability and likelihood to migrate or change employers. Men were certainly most mobile during the years in which their skills were in greatest demand, that is, between the ages of fifteen and nineteen, when they began work as putters, to about age forty, when they neared the end of their careers as hewers. Women, not surprisingly, were most likely to dominate in older communities at precisely that same age group, fifteen through forty, after men had secured work elsewhere. To give but one further illustration, Figure 4.1 plots the sex ratio in each age group in Houghton-le-Spring, an older community, and Coxhoe, a newly established mining area. A sex ratio of 100 indicates an equal number of men and women; a ratio above 100 indicates more males, below 100 a preponderance of females.

In Coxhoe, males outnumbered females in every age group below the age of fifty-five, an indication of the booming market for male labor. In Houghton-le-Spring, in contrast, women outnumbered men prominently

⁴¹ Ibid.

between the ages of fifteen and forty, the peak child-bearing years. Intuitively, one could also guess that given this distribution of men and women, fertility would be significantly higher in Coxhoe than in Houghton-le-Spring if only because there would be more extensive marriage opportunities available in the former. This would account for the fact that although Coxhoe only had about 12 percent more women of child-bearing age than Houghton (795 women between the ages of fifteen and forty-five as opposed to 710), the boomtown also had about 50 percent more children under five years of age (1,277 as opposed to 847).

To a great extent, Coxhoe's demographic profile was more akin to that of the mining region as a whole than that of Houghton-le-Spring. The work of Haines as well as that of Crafts has revealed that the Durham mining region was characterized by significantly higher fertility rates than the nation as a whole, as well as earlier and more extensive marriage.[42] That is, women married younger on average in Durham than elsewhere in England, and the percentage of women who "ever married" was the highest in Britain except for the women of Staffordshire. Moreover, women of the mining region of Durham had an average of one to two more children than the average British woman.

Migration patterns and the region's unique demographic profile can both be explained to a great extent by the segmentation of the labor market according to gender. While boys and men found a strong market for their labor in the collieries, women found employment in domestic service, in total the second largest field of employment in County Durham in 1841, or as dressmakers, milliners, agricultural laborers, schoolmistresses, and publicans or innkeepers.[43] Nonetheless, the participation of women in the work force in County Durham as a whole was far below that of the national average. In 1841, only 16.7 percent of those whose occupations were counted by the census enumerators were women compared to a national average of 26.4 percent.[44]

The sexual division of labor that restricted most women to the home in the coalfields existed long before the Victorians viewed women's work as a social problem and would remain long afterward.[45] Yet for women there

---

[42] Michael Haines, *Fertility and Occupation: Population Patterns in Industrialization* (New York: Academic, 1979), and N. F. R. Crafts, "Average Age at First Marriage for Women in Mid-Nineteenth-Century England and Wales: A Cross-Section Study," *Population Studies*, 32, no. 1 (1978), 21–5.

[43] *1841 Census: Occupation Abstract*, P.P. (1844), vol. 27, 42–3.

[44] Ibid., 297.

[45] Colls, *Pitmen of the Northern Coalfield*, 133–45. Norman Dennis, Fernando Henriques, and Clifford Slaughter's *Coal is Our Life* (1956; repr. London: Tavistock, 1969), 180–245, offers a classic account of women in the modern coal mining community. In general, see Brian

were glimpses of independence and brief escapes from work in the home that were cherished all the more. Local farmers, for example, employed women from the colliery villages at sowing and harvest times. "These two periods become such important eras in their lives," the Lambton colliery doctor wrote, "that it is common to hear the women date events by them."[46] Further, as Colls has shown, sexual segregation outside of the home and workplace was less rigorous than is often assumed.[47]

While the exclusion of women from work in the mines of the northern coalfield was accomplished during the first quarter of the eighteenth century,[48] within the next half-century, the development of new ventilation systems based on "coursing the air" and the extension of workings underground created a strong demand for child and adolescent labor. As early as age five or six, male children were employed in the Northeast as "trappers," or doorkeepers. They were relegated to small stone cubicles "about the size of a common fireplace,"[49] according to the mines subcommissioner of South Durham, where they operated doors designed to direct air currents through the pit and dissipate accumulations of dangerous gases. These doors often presented an obstacle to coal haulage, and a trapper's job was to open the doors when haulers needed to pass through and then to ensure that they were subsequently closed.

In the evidence collected by the 1842 Children's Employment Commission, a trapper's job was not deemed physically demanding, according to mid-nineteenth-century standards, but it was tediously boring. Trappers usually had to be at their stations by 4:00 A.M. in mines operating on a single shift. Before Tommy Hepburn's union, their days were fourteen to sixteen hours long, but Hepburn's union reduced that to twelve hours. On average, trappers earned about tenpence per day. Many boys testified that their loneliness and isolation were compounded by threats and violence from other workers as well as supervisors. One thirteen-year-old who admitted to falling asleep when left alone in the dark told the commissioners that he was hit by the overman and thumped by the haulers.[50] In this case, a sleeping trapper certainly threatened safety, production, and earnings; he thus bore the bruises of both management and labor.

Trappers could graduate to become "drivers" at about the age of twelve

Harrison, "Class and Gender in Modern British Labour History," *Past & Present*, no. 124 (August 1989), 121–158.

[46] *Reports from Commissioners: Children's Employment (Mines)*, P.P. (1842), "Appendix upon Education at Collieries," 730.

[47] Colls, *Pitmen of the Northern Coalfield*, 140–1.

[48] Flinn, *History of the British Coal Industry*, 333.

[49] *Reports from Commissioners: Children's Employment (Mines)*, P.P. (1842), vol. 16, 129.

[50] Ibid., 163. See also evidence on 158, 162–3, and 569.

or thirteen. A driver rode or pulled the ponies that hauled coal to the main shaft. Like trappers, the drivers worked twelve-hour days after Tommy Hepburn's union movement for which they earned about fifteenpence a day. Driving did not require great physical strength, but as Nicholas Wood, a coal owner and viewer, admitted, "The drivers are most liable to accidents."[51] Drivers frequently were thrown from their horses or fell from the trams hauling coal. John Percy, a ten-year-old driver at Percy Main Colliery, had already suffered two such accidents. In the first, the tram had run over his arm after he had fallen to the ground. In the second, his horse had stepped on his arm after he had been thrown. After each accident, he stayed off work for a month.[52]

Adolescent boys then succeeded to join the rank of putters at some time in their middle or late teens.[53] Putting was perhaps the most arduous task in mining for it entailed the manual haulage of coal in wheeled corves (or coal baskets) from the face to the shaftways. According to J. R. Liefchild, the children's employment commissioner, most corves averaged between six and eight hundredweight and, in Lord Durham's Littletown and Sherburn collieries, the daily distance traveled by a putter totaled about eight miles.[54] Putters were also introduced to the dominant form of wage payment in the collieries, the piece rate. George Elliot, the viewer of Monkwearmouth Colliery, believed the difficulty of putting necessitated piece rates to keep boys working. "Putting is extremely hard work," he explained, "but as putting is piece-work putters will work very hard, from the desire to earn increased wages."[55] In the 1830s and 1840s, putters were paid about 1s. 2d. to 1s. 4d. for every score of corves moved from the coal face. The size and extent of the workings of any one pit, however, meant that the distance of each put, the "renk," varied considerably. Therefore, each pit negotiated a list of putting rates. Typically, putters received a base rate per score for renks of between eighty and one hundred yards. Above a hundred yards, putters received an additional one pence per score for each twenty yards further.[56]

If putters worked alone, they could make up to four shillings a day, nearly equal to that of hewers. But often the work was too difficult for one boy

---

[51] Ibid., 587.

[52] Ibid., 578.

[53] The age at which workers advanced to the various stations of mining varied at each colliery. Walbottle Colliery, according to its viewer, Matthias Dunn, employed putters between the ages of fourteen and eighteen, while Percy Main Colliery testified its putters were aged seventeen through twenty-one. See ibid., 543–4, 557.

[54] Ibid., 544; *Report of the Commissioner... into the State of the Population in the Mining Districts, 1846*, P.P. (1846), vol. 9, 64.

[55] *Reports from Commissioners: Children's Employment (Mines)*, P.P. (1842), vol. 16, 642.

[56] Ibid., 586.

and it was common for an older boy, called the "headsman," to engage a younger boy, the "foal," as a helper. Headsmen, however, were the only boys paid by the firm and the foals were paid out of the earnings of their headsman. Two boys of approximately equal age and strength also could agree to share the burden of putting equally. In this case, the boys were referred to as "half-marrows," and they divided their earnings equally between them.[57]

While putters were first introduced to the adult forms of wage bargaining in the form of piece rates, they were also introduced to an adult form of job control, that of "drawing cavils." Cavilling was the process of drawing lots for workplaces, and it possessed the double advantage of fairly allocating easy and difficult working positions and protecting the putters, as well as the hewers, from favoritism or victimization.[58] Cavils were drawn and workplaces reallotted at regular intervals. It appears to have been common to draw cavils each fortnight, although this varied from colliery to colliery and sometimes from pit to pit. At Tyne Main Colliery, for example, putters' cavils were drawn monthly.[59] The principle, however, was to regularly give putters a fair opportunity to work an easier position and thus to increase their earnings.

Given the extent of child and adolescent labor in the mines, there is little doubt that the family played an important role in the workplace. Both coal owners and the Children's Employment Commission of the early 1840s placed special emphasis on the fact that collier families actively sought out employment for their children.[60] Mothers who sent their children down the pits because of the incapacitation of the male wage earner were frequently cited by the commissioners.[61] Similarly, the commissioners highlighted fathers whose earnings could not support the family without the additional income of child labor.[62]

Like textiles, in which the family structure appears to have been reproduced in the factory,[63] the collieries re-created the community at large and transmitted its values. Work in the mines was the primary medium through which the culture of patriarchy was articulated and elaborated. Fathers not only passed on their skills and experience to their sons, but they often bore

---

[57] Ibid., 522, 586.
[58] M. J. Daunton, "Down the Pit: Work in the Great Northern and South Wales Coalfields, 1870–1914," *Economic History Review*, 2nd ser., 34, no. 4 (1981), 586.
[59] *Reports from Commissioners: Children's Employment (Mines)*, P.P. (1842), vol. 16, 630.
[60] A. J. Heesom, "The Northern Coal-Owners and the Opposition to the Coal Mines Act of 1842," *International Review of Social History*, 25 (1980), 241–2.
[61] *Report from Commissioners: Children's Employment (Mines)*, P.P. (1842), 16, 585, 621, 630.
[62] Ibid., P.P., 162, 584, 618.
[63] Joyce, *Work Society and Politics*, 50–64; Reddy, *Rise of Market Culture*, 157–68.

the responsibility for disciplining their children as well.[64] Thus, the 1825 United Colliers union spoke tellingly of the effects of exploitation on their children, but also reserved to themselves the right to inflict corporal punishment upon those children who were lazy or impertinent.[65]

More importantly, the training of a trapper, driver, or putter was an indoctrination into the male world of work.[66] Jack Lawson, for example, recalled his first job as a trapper when he was twelve years old. He worked ten hours a day for tenpence, but his work in the mines entailed a "newly acquired status" in the family and community. "I was a man," he wrote, "and I knew it." He now was entitled to a new suit of clothes instead of hand-me-downs. He was accorded a larger portion of meat at dinner, stayed out at night as late as he wished, and, "even mother slightly deferred to me."[67] "Thus," Lawson concluded, "ten hours a day in the dark prison below really meant freedom for me."[68] Thomas Burt similarly recalled that "what he was most eager for in those early days was to grow up to ten years of age, when he could go down the pit"[69] Thomas Jordan, whose brief autobiography has survived, first went down the pits as a companion to his father who was an overman at Usworth Colliery.[70]

The initiation into the culture of patriarchy did not end with a boy's job as a trapper. Each rung on the occupational ladder to becoming a hewer was a further immersion in the male ethos and another step toward full participation in the community of men. Lawson's graduation to the ranks of the putters was an important aspect of masculinity: "To become a putter in a colliery like Boldon was to rank as a man."[71] Yet full status as a member of the community was only accorded to hewers. When Lawson became a hewer at age twenty-three, he recalled, "I was now a man." Perhaps sensing his own contradictory definition of the stages of manhood, he was moved to add that "a man is not really a man in Durham until he goes to the coalface."[72]

While workers may have tended to view child and adolescent labor as a rite of passage, it is readily apparent that coal owners viewed children's

---

[64] Williamson, *Class, Culture and Community*, 90.
[65] *A Voice from the Coal Mines*, 25–6.
[66] See the explication of the male-dominated mining community in Dennis, Henriques, and Slaughter, *Coal is Our Life*, 201–33. See also Martin Bulmer, "Social Structure and Social Change in the Twentieth-Century," in Martin Bulmer (ed.), *Mining and Social Change: Durham County in the Twentieth Century* (London: Croom Helm, 1978), 15–48.
[67] Jack Lawson, *A Man's Life* (London: Hodder & Stoughton, 1949), 74.
[68] Ibid.
[69] Watson, *A Great Labour Leader*, 33.
[70] John Burnett (ed.), *Useful Toil* (Harmondsworth: Penguin, 1977), 99–107.
[71] Lawson, *A Man's Life*, 94.
[72] Ibid., 121.

employment as an essential element of the creation and reproduction of the supply of labor. Buddle, like many others, believed that colliers could "never be recruited from an adult population."[73] Indeed, the furor aroused among the northern coal owners by Lord Ashley's 1842 bill to limit women and children's employment in the coal mines was predicated on their belief that "boys do not acquire those habits which are peculiarly necessary to enable them to perform their work in the mines" after age ten.[74] As Buddle explained to Hedworth Lambton, "We are decidedly of the opinion that if [boys] are not initiated before they are 13 or 14 – much less 16, 17, or 18 – *they will never become colliers.*[75]

Alternatively, it is unclear whether miners' trade unions recognized that their opposition to children's employment, apparent as early as 1825, might have had the effect of restricting the supply of labor. Both the United Colliers of 1825 and Hepburn's union of 1831–2 expressed their resistance in predominantly ethical terms; child labor was both inhumane and promoted ignorance.[76] Moreover, the question of child labor provided an initial stimulus to the organization of the Miners' Association in 1842.[77] However, the impact of the restriction of children's employment on the relative power of labor and capital in the market was never articulated in just those terms. Instead, as will be shown, evidence points to the influence of Methodist ideology in promoting an interest in child welfare and education among the miners.[78]

Therefore, the labor market, attenuated and segmented as it was, was of fundamental importance to the construction of community and social relations in the northern coalfield. Communities formed, matured, and declined based on the market for labor. Their social character, including their stability, gender relations, and even the presence of children reflected a particular juncture in the ebb and flow of workers through the community. Moreover, the patriarchy of the coal mining village was largely a function of the exclusion of women from the world of work and the male monopoly of labor outside of the home. The miners themselves, as we shall see, accepted the role of

---

[73] D.C.R.O., Londonderry Papers, D/Lo/C 142(1315), Buddle to Londonderry, 16 May 1842, quoted in A. J. Heesom, "The Northern Coal-Owners," 243.

[74] N.R.O., Coal Trade United Committee Minutes, 1840–44, 170–2, quoted in Heesom, "The Northern Coal-Owners," 243.

[75] D.C.R.O., National Coal Board Papers, I/JB/1786 quoted in Heesom, "The Northern Coal-Owners," 242.

[76] See *A Voice from the Coal Mines*, 35–6; "Meeting of the Pitmen," *Newcastle Chronicle*, 26 March 1831.

[77] Raymond Challinor and Brian Ripley, *The Miners' Association: A Trade Union in the Age of the Chartists* (London: Lawrence & Wishart, 1968), 209–13.

[78] See Chapters 7 and 8.

the market in their lives, and their struggles were largely premised on the attempt to free the market from the owners' control and to remove impediments to the labor market.

Paternalism and the market

This same struggle to secure power in the market was the background against which paternalistic authority was constructed in the first half of the nineteenth century. The social construction of order and authority in mining villages posed several potential problems to the coal industry. There was some concern, for example, that the separation of ownership from management threatened to break the traditional bonds of authority and thereby pose a threat to the social order. In this vein, Emile Cheysson, the late-nineteenth-century French industrialist, doubted whether firms that were not owner-operated could construct a paternalistic environment.[79] Yet as we have seen, the separation of ownership from management in the northern coal industry directed much social antagonism away from the owners of capital and focused it upon management. This shielded ownership from direct participation in labor relations and reinforced their social distance from labor.

If the direct power of management was therefore crucial in the construction of authority, the powers of the viewer and agent were also limited by several important restrictions. The viewer's powers over employment and housing were somewhat constrained by the institution of the bond, which, before 1844, secured both factors at least for one year. More importantly, the hewers' autonomy at the workplace, including the tradition of cavilling, significantly limited the viewer's ability to victimize and intimidate miners on the job. Thus employer control of the workplace, an essential component of the construction of deference in the northern textile factories,[80] was and remained relatively weak in the northern coal industry.

Nonetheless, the rapid expansion of coal production in the second quarter of the nineteenth century and the repeated bouts of unionization and labor unrest in the 1820s, 1830s, and 1840s led many of the largest coal owners to question implicitly the social effects of economic growth. Consequently, these years witnessed the attempts on the part of coal owners to erect a social order in the colliery village along paternalist and deferential lines. However, without greater authority at the point of production, the efforts of the coal owners were bound to be inconclusive. The respectable, sober miner

---

[79] Donald Reid, "Industrial Paternalism: Discourse and Practice in Nineteenth-Century French Mining and Metallurgy," *Comparative Studies in Society and History*, 27, no. 4 (1985), 593. I would like to thank Professor Reid for making a copy of his article available to me.

[80] Joyce, *Work, Society and Politics*, 96–103.

was less the social construct of the employer's paternalist endeavors than the creation of a segment of the mining community itself.

The coal owners' efforts in the colliery villages were directed to three prinicipal areas: education, religion, and benefit societies. We now know a good deal about the provision of education in the first half of the nineteenth century.[81] It is clear that the collieries were generally well supplied with a variety of dame's schools, day and night schools, as well as Sunday schools.[82] Moreover, as Laqueur has shown for Sunday schools in particular,[83] these schools were, in the words of the mines commissioner, institutions where "colliers appear in general to prefer sending their children" in part because they were "kept by men of their own class."[84] Also, Brendan Duffy's work clearly points to the growing role of the established Church and other denominations in providing education in the colliery districts, a role that probably outweighed that of company-sponsored schools.[85] Finally, the chronology of the establishment of public elementary schools of various denominations in the northern coalfield clearly indicates a crescendo of school building that peaked between 1840 and 1850 and subsided thereafter[86] (Figure 4.2).

Denominational motives for school building, however, differed somewhat from those of the coal owners. The established Church, according to Duffy, provided educational facilities particularly in order to counter the influence of Methodism.[87] Thus, the chronology of the establishment of public schools correlates with the Church of England's realization of its weakness in the face of nonconformity. Still this was not the only reason the Church built schools. In addition, they were eager to stabilize the social order in the expanding colliery villages of the Northeast, and it was here that Church motives coincided with those of the coal owners.[88]

---

[81] See Ray Pallister, "Educational Investment by Industrialists in the Early Part of the Nineteenth Century in County Durham," *Durham University Journal*, n.s., 30 (1968–9), 32–8; Robert Colls, " 'Oh Happy English Children!': Coal, Class and Education in the North-East," *Past & Present*, no. 73 (November 1976), 75–99; A. J. Heesom and Brendan Duffy, "Debate: Coal, Class and Education in the North-East," *Past & Present*, no. 90 (February 1981), 136–51; Robert Colls, "Debate: Coal, Class and Education in the North-East: A Rejoinder," *Past & Present*, no. 90 (February 1981), 152–65.

[82] *Report of the Commissioner... into the State of the Population of the Mining Districts, 1846*, P.P. (1846), vol. 24, 13–14.

[83] Thomas W. Laqueur, *Religion and Respectability: Sunday Schools and Working Class Culture, 1780–1850* (New Haven, Conn.: Yale University Press, 1976).

[84] *Report of the Commissioner... into the State of the Population of the Mining Districts, 1846*, P.P. (1846), vol. 24, 13.

[85] Duffy, "Coal, Class and Education," 146.

[86] Ibid., 142–51.

[87] Ibid., 144–5.

[88] Ibid., 145–6.

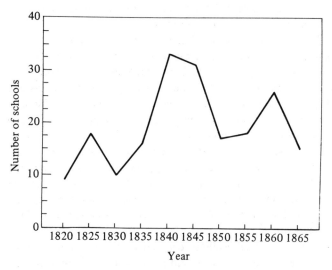

Figure 4.2. Public elementary school building, 1820–69. (From Brendan Duffy, "Debate: Coal, Class and Education in the North-East," *Past & Present*, no. 90 [February 1981], 143.)

There is little doubt that coal owners increasingly viewed education, religion, and benefit societies as important means of securing a stable, as well as a deferential, work force. Both the 1831–2 and 1844 miners' strikes evoked demands and concerns for the education of the mining population on the part of employers.[89] Colls argues correctly that coal owners expressed little interest in education per se, rather only in schooling that was provided under their influence and largesse.[90] This common concern was expressed to the mines commissioner in 1846 by one "intelligent person" who reported: "The present kind of education is inadequate. We have had schools of one kind and another for a long time, but they have not prevented the people from being deluded when attempts have been made upon them, as at the time of the strike."[91] Similarly, the 1831–2 strike provoked Lord Durham to establish the Lambton Collieries Benefit Society to encourage the "habits of industry, sobriety, and religion" and to contribute an additional one-sixth of the total funds raised by voluntary subscription.[92] These funds, however,

[89] On the 1831–2 strike, see Duffy, "Coal, Class and Education," 146–7; on 1844, see Colls, " 'Oh Happy English Children,' " 91–3.
[90] Colls, " 'Oh Happy English Children,' " 96.
[91] *Report of the Commissioner... into the State of the Population in the Mining District, 1846*, P.P. (1846) vol. 24, 27.
[92] Duffy, "Coal, Class and Education," 147–8; Holland, *History of Fossil Fuel*, 301–2; *Reports from Commissioners: Children's Employment (Mines)*, P.P. (1842), vol. 14, 723–4.

were not controlled solely by the men and thus could not be used to support striking miners as had occurred during 1831.[93]

Still, while the mines commissioner, for one, exhibited a concern bordering on paranoia with the manner in which working-class attitudes were formed independent of the upper class's "affectionate tutelage,"[94] the principal goal of industrial paternalism was not necessarily that of ideological domination. Instead, it was the creation of stable colliery communities insulated from the forces of the labor market. The 1846 parliamentary inquiry into education in the mining districts points directly to the coal owners' interest in paternalism as an investment that was supposed to yield a return in the form of a stable, immobile work force. In fact, perhaps the dominant theme running through the evidence from coal owners and their agents is not their drive to achieve ideological hegemony, but their assumption that paternalism lowered costs and their palpable surprise at the subsequent failure of their endeavors to subvert the market.

An agent of Haswell Colliery claimed that "our men are less given to change than at most collieries," and proudly listed the reasons why: steady work, good houses, ample gardens, efficient village drainage, low incidence of disease, and cleanliness ("We keep everything clean about them, *carting away all the ashes, &c.*").[95] Similarly, one of Lord Londonderry's agents listed the marquess's previous efforts: provision of houses and gardens, free coals in the case of an accident or injury, a school at Pensher for the past forty years and a library for the past ten years, regular subscriptions to the national school and private schools at West Rainton and Pittington, support for local Sunday schools, and so forth. Still, George Elliot, a viewer at the collieries, recalled that despite the fact that the men "had no grievances to complain of, they 'struck' with the rest in April 1844."[96] One of his chief concerns was the cost of the strike:

> It cost us 30s. per man to bring the men from Ireland [to break the strike]; then we gave them 3s. a-day and their food for five months, and they were so awkward at the work at first that they could scarcely earn what their food cost us. Then there was the expense of the military and police; the loss of trade; all the way-leaves, rents, salaries, and all the other standing expenses which went on all the time the men were idle. Altogether, the strike was the cause of great expense and loss to our collieries.[97]

---

[93] On the 1831–2 strike generally, see Chapters 7 and 8.

[94] The term is from J. S. Mill, *Principles of Political Economy*, vol. 2 (Boston, 1848), 319–20, cited in Dutton and King, "The Limits of Paternalism," 72.

[95] *Report of the Commissioner... into the State of the Population in the Mining Districts, 1846*, P.P. (1846), vol. 24, 19.

[96] Ibid., 15.

[97] Ibid., 16.

Elliot concluded that stricter control of the labor market, achieved through the substitution of monthly for yearly bonds after the strike, was of great importance in improving the conduct of the workers. Rather than emphasize the value of ideological control, Elliot explained that the monthly contract "is working an improvement in the morals of the people, because they now see that they can be discharged readily for any misconduct."[98]

The hidden paternalism that mattered most widely, as Patrick Joyce put it, also was employed throughout the collieries.[99] Whereas housing, education, and employers' benefit societies were institutional embodiments of paternalism, the miners' lives and those of their families were penetrated in a hundred smaller ways by their employers. In the northern coalfield, for example, colliery doctors increasingly played an important role in the provision of medical care even though their care often was at odds with local remedies.[100] Colliery agents organized subscriptions for the families of victims of explosions and accidents, as in the case of the famous Haswell Colliery disaster of 1844. Many collieries provided transportation or discounted fares to local market towns. Lord Durham's and Lord Ravensworth's collieries, for example, both provided horse-drawn carts "to bring home for [the miners] what they have bought at market" every Saturday.[101] Haswell Colliery made arrangements to secure half-price tickets on the railway to Sunderland, where the workmen and their families went to market.[102] At Coxlodge Colliery, after the bonds were signed in 1831, 128 pitmen's wives were invited to tea and a dance at the home of colliery agent.[103]

These efforts followed a pattern designed not only to persuade the labor force, but more importantly to secure a constant labor supply to the collieries.[104] Embedded in the language and custom of paternalism is the expression of a functional antimarket ideology. Many coal owners believed that the workers had willfully severed their customary social ties through a combination of ignorance and misdirection. However, they refused to recognize the fact that the restoration of these ties would also serve to undermine an active labor market. In a typical response, after the 1831–2 strike, R. W. Brandling, chair-

---

[98] Ibid., 15–16, also quoted in Colls, *Pitmen of the Northern Coalfield*, 70.
[99] Joyce, *Work Society and Politics*, 145.
[100] *Reports from Commissioners: Children's Employment (Mines)*, P.P. (1842), vol. 16, 727–8.
[101] *Report of the Commissioner . . . into the State of the Population of the Mining Districts, 1846*, P.P. (1846), vol. 24, 23, 63.
[102] Ibid., 64.
[103] *Newcastle Chronicle*, 18 June 1831.
[104] Peter Stearns has argued that labor shortages constituted one of the earliest reasons for the development of paternalism in French industry; see Stearns, *Paths to Authority: The Middle Class and the Industrial Labor Force, 1820–48* (Urbana, Ill.: University of Illinois Press, 1978); Pallister, "Educational Investment by Industrialists," 33; Reid, "Industrial Paternalism," 582–5.

man of the coal owners' cartel, hoped that the men he had engaged as strike breakers would be "of more upright principles and with more correct notions of the rights and relative duties of Masters and Servants."[105]

Both Lord and Lady Londonderry certainly viewed their efforts to provide education as a function of industrial good lordship. Lord Londonderry hoped that the provision of three new schools at Pensher, Rainton, and Pittington collieries after the 1844 strike was a means of proving that intention. Yet again, ideological hegemony was less important than subverting the labor market. Similarly, he employed the same language of paternalism to avoid contributing to a public subscription to aid the victims of the Haswell Colliery explosion in 1844: "In proportion as the Collier or Pitman devotes his labour and runs the risk of the Mines for the benefit of the Employer so is the latter in common duty, honesty and charity bound to provide for and protect those who are bereft of their protection by any fatality that occurs."[106] Lady Londonderry later exhibited similar patterns of thoughts in resistance to the formation of a £50,000 strike indemnity fund for all collieries in 1854: "My people have advantages that all have not," she wrote. "They are cared for, their comfort looked to, and the education of their children watched over, and I would fain hope they will be kept straight by gratitude and good feeling as by their own good sense."[107]

Interestingly, this new industrial paternalism of the 1830s and 1840s also served to illuminate the weaknesses and failures of the gentry and clerisy in the mining regions of the Northeast.[108] J. R. Leifchild, the children's employment commissioner, bemoaned the fact that the "active benevolence of the higher ranks which induces them to visit the habitations of the working classes; to counsel, guide, and instruct them; to patronize their schools, and encourage their attempts at order, frugality, and amelioration, are here wholly deficient."[109] In 1830, when the government suggested that special yeomanry corps be established in regions of dense concentrations of population and valuable machinery, Henry Morton, the Lambton estate agent, lamented that "there are few resident Gentlemen in this county to chuse [sic] upon and still fewer proper for such a crisis."[110] The Church's renewed efforts

[105] Quoted in Duffy, "Coal, Class and Education," 147.
[106] D.C.R.O., Londonderry Papers, D/Lo/C 739(1), Londonderry to Day, 15 October 1844, also quoted in A. J. Heesom, "Entrepreneurial Paternalism: The Third Lord Londonderry (1778–1854) and the Coal Trade," *Durham University Journal*, n.s., 35, no. 3 (1974), 245.
[107] D.C.R.O., Londonderry Papers, D/Lo/B 6(2), n.d., (c. 1854).
[108] On the continuing weakness of the Durham gentry in the later nineteenth century, see T. J. Nossiter, *Influence, Opinion and Political Idioms in Reformed England: Case Studies from the North-East, 1832–1874* (Hassocks: Harvester, 1974) 50–1.
[109] *Reports from Commissioners: Children's Employment (Mines)*, P.P. (1842), vol. 16, 519.
[110] Lambton Mss., Morton to Durham, 6 December 1830; D.C.R.O., Londonderry Papers, D/Lo/C 86(10), Durham to Londonderry, 8 December 1830; D/Lo/C 142(628), Buddle to Morton, 10 December 1830.

in the Northeast during the 1840s, according to Duffy, also was part of an effort "to compensate for the absence of a resident gentry."[111] Church officials acknowledged that the populations of the collieries were "rapidly collected together without any of that mixture of rank, and intercourse between the rich and poor, which is so beneficially exercised in most parts of the kingdom."[112] The problem was viewed to be so chronic in some quarters that specific gentlemen and clerics, including the notorious pluralist Henry Phillpotts, were attacked in the local press for neglect of their social responsibilities.[113]

The gentry, nevertheless, defended their role in local society by arguing that they functioned as intermediaries between capital and labor.[114] During the early months of Tommy Hepburn's strike in 1831, Lord Brougham made the suggestion that colliery viewers and agents might make acceptable J.P.s. The bishop of Durham, who exerted a significant influence over the county magistracy through clerical J.P.s. who held rectories in his patronage,[115] promptly declared that "Agents, especially in the mining districts, were not proper persons to act as Justices."[116] James Losh, a prominent north country attorney, small colliery owner, and friend of Lord Brougham, warned the Lord Chancellor of "numerous resignations or a general remonstrance from the Bench" if such a plan were implemented in Durham. He further advised Brougham that "Colliery Viewers will neither be acceptable (as brethren) to the present Magistrates nor in fact proper men for Justices of the Peace."[117]

The magistrates on the benches of the Quarter and Petty Sessions tended to be more uniformly of the gentry class in Durham than in the Tyneside areas of Northumberland, where coal owners acted frequently as J.P.s. This is not surprising given the fact that Tyneside coal firms were on a generally smaller scale than those south of the Tyne. In Northumberland, coal owners such as R. W. Brandling, chairman of the coal owners' cartel, T. W. Beaumont, Charles Bigge, and Sir Matthew Ridley all served as magistrates.[118] In Durham, however, coal owners rarely served on the bench in the early

---

[111] Duffy, "Coal, Class and Education," 145.

[112] Quoted in ibid., 145–6.

[113] Fordyce, *History and Antiquities*, vol. 1, 651; William Lorraine, a Houghton-le-Spring magistrate, was condemned for his failure to reside in the localities in the *Newcastle Courant*, 25 June 1831.

[114] Norman McCord, "The Government of Tyneside, 1800–1850," *Transactions of the Royal Historical Society*, 5th ser., 20 (1970).

[115] See Jaffe, "Economy and Community," 135–7.

[116] E. Hughes (ed.), *The Diaries and Correspondence of James Losh* (Newcastle-upon-Tyne: Surtees Society Publication no. 174, 1963), vol. 2, 192.

[117] Ibid., 190.

[118] William Parson and William White, *History, Directory, and Gazetteer of the Counties of Durham and Northumberland* (Leeds, 1827–8), vol. 2, 69–70.

nineteenth century, except for Lord Londonderry, whose heavy hand reached as far as his arm could stretch.

Nevertheless, in both Durham and Northumberland, magistrates took pride in their landed interests and their semblance of independence between capital and labor. In the 1831–2 miners' strike, eleven Northumberland magistrates offered to mediate the dispute between the miners and the coal owners. In a notice published in the *Newcastle Chronicle*, they indeed argued that they could be trusted to act in a disinterested manner specifically because they had no connection with the coal trade.[119] During the same strike, Durham magistrates refused to execute warrants for the eviction of union leaders from their company homes after a temporary settlement had been reached.[120]

The workers themselves often accepted the mediation of the magistracy and abided by their judgments. In June 1831, the miners of Black Boy Colliery complained that the baskets used to measure piece rates, the corves, contained more than the standard measure of thirty-three gallons. Two magistrates, John Trotter and Robert Surtees, measured three corves selected mutually by the men and the owners and found that they held thirteen gallons more than the standard measure. Trotter and Surtees decided that the miners deserved an additional 1s. 4d. for every two corves delivered to the bank in compensation for the extra measure.[121]

Actions such as these have led Norman McCord to conclude that the conciliatory activity of local magistrates was "more representative of British society...than any of the more startling manifestations of oppression or disaffection."[122] To a certain extent this is true. However, the gentry and magistracy were also painfully aware of the fact that a disruption in the coal trade threatened not only the coal interests in Durham and Northumberland, but all the subsidiary trades including shipping and iron. A mass disruption of the coal trade ultimately threatened all property. Thus the magistracy acted not only to mediate between owners and miners, but to protect themselves as well.

The peripheral role of the gentry in industrial relations should not obscure the nature of the struggle. The autonomy of the worker at the point of production shifted the frontier of control off of the shopfloor and into the community. Control of the labor market, through the annual bonds or the cultivation of the ties of paternalism, became, along with the Vend, one of the principal means by which coal owners could influence

[119] *Newcastle Chronicle*, 7 April 1831.
[120] Lambton Mss., Morton to Durham, 19 May 1831.
[121] D.C.R.O., T. Y. Hall Papers, NCB I/TH/20(9).
[122] McCord, "Government of Tyneside," 28.

the terms of exchange between classes. For their part, workers relied upon workplace bargaining and their ability to restrict production to achieve the same goal. It is to this side of early industrial market relations to which we now turn.

5

Work and the ideology of
the market

If all kinds of labour were perfectly free, if no unfounded prejudice invested some parts, and perhaps the least useful, of the social task with great honour, while other parts are very improperly branded with disgrace, there would be no difficulty on this point, and the wages of individual labour would be justly settled by what Dr Smith calls the "higgling of the market."

Thomas Hodgskin, *Labour Defended against
the Claims of Capital* (1825)

In September 1803, in preparation for the annual signing of the bonds, the miners at Lord Delaval's Hartley Colliery on Tyneside petitioned their viewer for an increase in piece-work rates.[1] At first, the Delaval agent, Paul Forster, hoped to resist the miners' demands, but when he read to them the terms of their bond "the general cry was we must have all that is in the petition." In order to secure the necessary complement of hewers to work the colliery, he soon proposed a very modest list of piece-rate increases that, to his dismay, was rejected by all but two miners. In the meantime, three men and a boy had left the colliery and were bound to another. Into the second week of October, only one more hewer had agreed to the rates offered by Forster. The vast majority of the men continued to hold out, according to Forster, "supposing that there would be a great scarcity of men as realy [sic] is the case."

On 9 October, the workings of the regional labor market for hewers began to threaten the future operation of the Delaval colliery. On that day, seven men and two boys were hired away from Delaval to work elsewhere. More importantly, perhaps, a man from Hebron [Hebburn] Colliery came amongst the Delaval pitmen "telling of the great earnings that was there, which inflam'd the peoples minds to a great hight [sic]." To protect the work force of his Lordship's colliery, Forster had to act quickly. This time he offered significant piece-rate increases; in one case, double that of the original offer.

---

[1] N.R.O., Delaval Papers, ZDE 6/5/7, Forster to Delaval, 17 October 1803. See also Colls, *Pitmen of the Northern Coalfield,* 76–7.

When the men still appeared reluctant to bind for the coming year, Forster's supervisor, John Bryers, sweetened the offer further by including "a Little hand money" of up to £2 10s. each, which finally secured the binding of the men.[2]

This glimpse of miners' activism at Hartley Colliery in the opening years of the nineteenth century illustrates the degree to which informal collective bargaining constituted the accepted terrain of local labor relations. These actions were similar, if not identical, to the modes of behavior followed by miners and their unions throughout the early nineteenth century: the presentation of petitions to the viewer or agent demanding revisions in piecerate schedules, followed by some form of collective bargaining, and then the organized cessation of labor if need be. This particular form of bargaining, as we have seen, was in part an aspect of the rise of managerial levels within the firm, a development that also had the tendency to extract ownership from the cash nexus of employer and employee.[3] More importantly, however, bargaining was rooted in the structure of work, particularly in the manner in which the duality of piece rates as both a method of exploitation and a symbol of independence mirrored the ambiguous social position of the miner between common laborer and commodity producer. The miners' working lives were an illustration of this fundamental and profound contradiction embodied in the piece-rate system.

Bargaining and piece rates also served to link miners to the mechanics of the market. They thus established the foundation for the creation of the late-nineteenth-century system of industrial relations that has often been characterized as economistic and accommodationist. Moreover, wage bargaining, and the concomitant efforts to control both output and the movement of labor, constituted one of the ways in which labor sought to influence market relations. As such, it was one of the principal sources of working-class power in early industrial society.

Therefore, labor activism in the mines was not premised upon the struggle for control of the labor process or resistance against the "real subordination" of labor.[4] Instead, the relative autonomy of the miner necessitated the relocation of exploitation, and thereby the frontier of control, to the product

---

[2] N.R.O., Delaval Papers, ZDE 6/5/7, Forster to Delaval, 17 October 1803.
[3] See Chapter 3.
[4] Price, "Labour Process and Labour History," 59–75. See also the subsequent debate: Patrick Joyce, "Labour, Capital and Compromise: A Response to Richard Price," *Social History*, 9, no. 1 (1984), 67–76; Richard Price, "Conflict and Co-operation: A Reply to Patrick Joyce," idid., no. 2 (1984), 217–24; and, Patrick Joyce, "Languages of Reciprocity and Conflict: A Further Response to Richard Price," ibid., 225–31. The *locus classicus* of modern historical interpretations of the "real subordination" of labor under capitalism is Harry Braverman, *Labor and Monopoly Capital* (New York: Monthly Review Press, 1974).

and labor markets.[5] It is for this reason that union movements among the miners during this era expressed a desire to protect workplace bargaining, particularly through the institution of restrictions on output and the control of the movement of labor.[6] The argument presented here does not deny the general analytical validity of the conception of workplace relations as based upon a "structured antagonism" between management and labor,[7] but it is intended to suggest that the control of market relations as opposed to productions relations, was the arena in which social relations were defined in the early nineteenth century.

The structure of work

A hewer's work required great daring, strength, and skill. J. R. Harris has remarked aptly that hewing, like other crafts, was based upon "a precarious combination of manipulative skill embodying a physical training and a judgment requiring both experience and intelligence."[8] This special combination of expertise and experience constituted the "knacks" of the trade. The knacks of hewing included the methods of carving and cutting down coal from the face, the ability to distinguish grades and quality of coal, and the experience, bordering on intuition, to recognize the impending dangers of accumulating gas or the threatening "creep" of settling walls and roof.[9]

The mysteries of the hewer's craft, however, were not so impenetrable to the uninitiated so as to function as a form of protection from the competition of other laborers. Royden Harrison has emphasized the fact that a hewer's skills were real but also expendable. Inexperienced labor might be less efficient and produce an inferior product, but they could produce a product nonetheless. Professor Harrison certainly was correct to draw attention to the threat posed by competition from below, for the employers' most potent weapon against union activity in the Northeast always was the importation of blackleg workers to break strikes.[10]

[5] See Craig R. Littler and Graeme Salaman, "Bravermania and Beyond: Recent Theories of the Labour Process," *Sociology*, 16, no. 2 (1982), 251–69, who suggest that capitalism is principally driven by the desire to accumulate and that strategies of control away from the point of production therefore merit closer attention.

[6] See Chapters 7 and 8.

[7] Edwards, *Conflict at Work*, 5–6.

[8] J. R. Harris, "Skills, Coal and British Industry in the Eighteenth Century," *History*, 61, no. 202 (1976), 182.

[9] Flinn, *History of the British Coal Industry*, 91–2, 339–40; Colls, *Pitmen of the Northern Coalfield*, 11–33; the skills of the Scottish hewer are admirably described in Alan Campbell and Fred Reid, "The Independent Collier in Scotland," in Royden Harrison (ed.), *Independent Collier: The Coal Miner as Archetypal Proletarian Reconsidered* (New York: St. Martin's, 1978), 54–61.

[10] Harrison, *Independent Collier*, 4–7.

It has long been recognized that a miner's work entailed an extraordinary degree of autonomy and independence at the workplace.[11] Arthur Young expressed some surprise at the fact that the miners were "greatly impatient of control" while on his tour of the north of England in the 1760s.[12] Hoping to explain this phenomenon, Carter Goodrich, the twentieth-century labor economist, claimed that the miner's freedom and autonomy were a function of the technique of coal extraction and the geography of the workplace.[13] Thus, he argued, the bord-and-pillar method of working coal (which predominated in Durham and Northumberland) allocated separate "stalls" to individuals or work groups that simultaneously prevented effective supervision or management while inculcating a sense of independence and job control.[14]

The geography of production, however, was not the only source of miners' freedom. Autonomy at the workplace was further reinforced by the practice of cavilling, that is, the drawing of lots at regular intervals to redistribute workplaces in the mine. Cavilling, as Daunton has shown, varied in practice from pit to pit and from colliery to colliery, but it functioned to equalize the chances of working difficult places, which translated into lower earnings on the piece-rate system, as well as to protect the miners from both victimization and favoritism.[15]

Additionally, the "marra" system generally has been invoked as further evidence of the miner's autonomy at the site of production. The marra was the hewer's mate who worked the same stall on alternate shifts down the pit. Marras epitomized another aspect of the miner's autonomy underground, for they chose to work the stalls together, they cavilled together, they alternated shifts usually each week, and relied on one another's work habits and practices to maintain their earnings and ensure their safety.[16] However, it

---

[11] Carter Goodrich, *The Miner's Freedom* (Boston: Marshall Jones, 1925); more recently, see M. J. Daunton, "Down the Pit," 578–97.

[12] Arthur Young, *Northern Tour* (1768), vol. 2, 261, quoted in Goodrich, *Miner's Freedom*, 17.

[13] Goodrich, *Miner's Freedom*, 19.

[14] Daunton, however, has argued that the technique of coal extraction may have been less significant than it at first appears. His comparison of the bord-and-pillar methods used in the Northeast with the long-wall technique of South Wales accounts for differences between the two regions with reference to the division of labor within the mines rather than to the precise method of extraction. Nevertheless, Daunton maintains that before mechanization both techniques of extraction were characterized by significant degrees of autonomy and freedom from supervision. See Daunton, "Down the Pit," 580–5.

[15] Daunton, "Down the Pit," 585–6. The situation in France, as Donald Reid has shown, was significantly different. There, by the late nineteenth century, foremen distributed work places and consequently possessed a greater degree of control over the mining labor force. See Reid, "Industrial Paternalism," 593–4.

[16] Daunton, "Down the Pit," 587–8; Bill Williamson, *Class, Culture and Community*, 89–90.

is unclear exactly how prevalent the marra system was among hewers before midcentury. Generally, the double-shift system of working mines, which was a necessary precondition to the creation of marras among hewers, does not seem to have been common before the 1840s,[17] although the shift system in the Northeast was certainly not unknown before that time: Dunn speaks of a "night shift" at one of Hetton Colliery's pits in 1831,[18] Buddle planned for two shifts at Elswick Colliery in the first decade of the nineteenth century,[19] and two shifts apparently were down Felling Colliery when disaster struck in 1812.[20] Still, when Buddle calculated the productive capacity of the coal industry for a parliamentary committee in 1836, he based his estimates upon 260 working days per annum all on single shifts,[21] an indication that double-shift work was not common practice. Nevertheless, it might be noted that marras did function commonly as a work group among the putters who needed extra assistance to push the heavy coal wagons. John Holland, author of a mid-nineteenth-century treatise on the coal trade, associated marras only with putters and cited the well-known "Collier's Song" that begins, "As me and my marrow were putting our tram."[22]

The miners' autonomy at the workplace was not solely predicated on their conscious control of the point of production. In part, the miners' freedom also was a function of the unwillingness of management to secure greater control of the workplace. Matthias Dunn, an early viewer of Hetton Colliery and later one of Her Majesty's inspectors of mines, saw virtue in the fact that in the bord-and-pillar system, "each workman is provided with an independant [sic] working place, clear of the confusion and inconvenience attending the *longwall* system."[23]

Management elected to avoid this "inconvenience" by exercising its most significant form of control over production through the terms elaborated in the annual contract, or bond.[24] The bonds regulated the imposition and rate

[17] Colls, *Pitmen of the Northern Coalfield*, 27–8. The 1844 miners' strike apparently precipitated the institution of the shift system in some collieries. Hetton, Elemore, and Appleton collieries, for example, adopted shifts in order to retain the strike breakers they had trained during the dispute and to reemploy their former complement of experienced hewers. In effect, the shift system put the larger workforce on short time. See *Report of the Commissioner ... into the State of the Population in the Mining Districts, 1846*, P.P. (1846), vol. 24, 20.

[18] N.C.L., Matthias Dunn's Diary, 8 December 1831.

[19] Flinn, *History of the British Coal Industry*, 368.

[20] P. E. H. Hair (ed.), *Coals on Rails*, 6.

[21] *Report of the Select Committee ... into the State of the Coal Trade*, P.P. (1836), vol. 11, 118.

[22] Holland, *History of Fossil Fuel*, 294.

[23] Matthias Dunn, *A Treatise on the Winning and Working of Collieries* (Newcastle, 1848), 135, quoted in Daunton, "Down the Pit," 583.

[24] See Chapter 3.

of fines for poor-quality work, the forms and standards of the measurement of production, and certain standards of work such as the width or height of headways and bords.[25] Thus, except for the measurement of headways and bords that were necessary for the structural support of the mine, management and ownership relinquished its control of the actual productive process and exerted authority only through and over the finished product. In this sense, coal mining in the Northeast was characterized by a dysfunction between a relatively advanced capitalist superstructure and a form of labor organization and production more appropriate to "manufactures" or the workshop.[26]

However, the bond also imposed certain limits on the power of management through the temporary protection it afforded labor from victimization and short-term cyclical fluctuations. Viewers do indeed seem to have followed the terms of the bond regarding the annual employment of labor, and once bound to a colliery, a miner could feel that his employment, if not his earnings, was secure for the coming year. Henry Morton, Lord Durham's agent, explained that while he could dismiss "deputies," who were lower-ranking supervisory personnel, with impunity, he had to wait until "next year [to] have an opportunity of proscribing all those infamous Rogues, the Ranter Preachers – and other bad characters."[27]

Management's prerogatives were often limited by the bond in one other important area, that of industrial relations. Once the bonds had been accepted, it was largely true that "the local government of a colliery, as it refers to fines, is purely despotic," as the United Colliers protested in 1826.[28] However, after 1810, the bonds did include a recognized system of arbitration to settle several areas of dispute that arose between management and labor, particularly in regard to disputes concerning the measurement of production and the negotiation of piece-work prices.

In the first area, that of the measurement of production, piece-rate payments in this era were made according to the volume of coal as measured in corves, or coal baskets, sent to the surface. A common complaint among miners was that corves contained much larger measure than the standard of twenty pecks: "How galling is the consideration that though we are bound

---

[25] See, e.g., the terms of the bond published by the United Colliers Association in *A Candid Appeal to the Coal Owners and Viewers of the Collieries on the Tyne and Wear* (Newcastle-upon-Tyne, 1826); and Hytton Scott, "The Miners' Bond in Northumberland and Durham," pt. 1, *Proceedings of the Society of Antiquaries of Newcastle-upon-Tyne*, 11: no. 2 (1947), 55–78.

[26] On the theories of manufacturing and domestic production, see Berg, *Age of Manufactures*, 69–77.

[27] Lambton Mss., Morton to Durham, 22 May 1832.

[28] *A Candid Appeal*, 11.

to send 20 pecks in our corf to bank, yet when we know that our corf contains
22 pecks and 7 quarts," the United Colliers of 1825 complained.[29] If the
measurement of corves was disputed, the bond allowed miners to be present
at the measurement of the corves. The hewers, however, were obligated to
give the viewer notice, in some cases several hours and in others several
days, before the corves were measured, and they were not recompensed for
any extra measure sent to the bank if the corves were found to be too large.
Both the United Colliers union of 1825 and Tommy Hepburn's union of
1831–2 accepted in principle this form of settling disputes but demanded
payment for overmeasure and a reduction of the number of days' notice
required before inspection.[30]

More significantly, the bonds included a general provision for resolving
all industrial disputes that were not specifically covered by the terms elaborated within them. In such instances, both the miners and owners possessed
the right to appoint a qualified viewer to arbitrate the case. If the two
appointed viewers could not reach an agreement, then they were expected
to appoint a third viewer to act as an umpire. The decision of the umpire
was deemed to be "conclusive."[31]

These negotiating apparatuses of the bonds were never strictly adhered
to by either labor or capital. There is evidence that local disputes over
measurements, for example, were settled outside the strictures of the bond
and in a more ad hoc fashion. The miners of Black Boy Colliery in 1831,
as noted, engaged the mediation of local J.P.s, who found for the hewers
and ordered suitable recompense for their extra output.[32] The miners at
Lord Londonderry's Rainton and Pittington collieries simply sent corves to
the bank that were not fully loaded and insisted upon payment for a full
corve.[33] However, this last case was clearly extraordinary; such direct action
occurred less commonly than protests that were followed by some form of
mediation or conciliation.

In the matter of disputes concerning piece-work prices, both parties had
recourse to the process whereby two viewers were selected from outside
collieries to act as arbitrators. Upon failing to reach a settlement, they then
could select jointly a third viewer, who had the power to impose a settlement
upon both parties. According to the bonds, the umpire's judgment was final
and could not be appealed.[34] Whether and how frequently this clause of the

[29] *A Voice from the Coal Mines*, 11.
[30] *A Candid Appeal*, 15; on Hepburn's union, see Chapter 7.
[31] See, e.g., the bond for 1812 at Washington Colliery reprinted in Scott, "The Miners' Bond,"
     pt. 2, 87–90, and *A Candid Appeal*, 9.
[32] D.C.R.O., T. Y. Hall Papers, NCB I/TH/20(9). See Chapter 4.
[33] D.C.R.O., Londonderry Papers, D/Lo/C 142(753), Buddle to Londonderry, 8 July 1831.
[34] See the copy of the bond published by the United Colliers in 1826 in *A Candid Appeal*, 9.

bond was strictly invoked, as in the case of the measurement of production, is a matter of some question. At the Lambton collieries in 1810, the system seems to have worked smoothly and equitably. There, upon the suggestion of the hewers, two viewers arbitrated the case for the owners and the men. Their judgment was deemed acceptable by both parties and accepted.[35] However, evidence from Waldridge Colliery in 1831 points to the fact that the viewer there was sufficiently unfamiliar with this system as to suggest the formation of an arbitration panel comprised of several viewers from more than a half-dozen collieries.[36] In the same dispute, both parties finally did appoint two viewers, but their settlement was not accepted by the miners, and the final stage of binding arbitration never was reached. In yet another case, the colliers at Whitefield and Townley Main Colliery were simply not rehired when they petitioned for increases and were replaced by blacklegs.[37] However, it is likely that the experience of the hewers at Hetton Colliery and Burdon Main Colliery was the most common; there, they bargained directly with their viewers and eventually reached a settlement without recourse to outside mediators or arbitration.[38]

There is, nonetheless, good reason to believe that the miners accepted the principle, if not the precise form, of the mediation and arbitration clauses of the bond. The United Colliers argued that arbitration was not only an equitable principle, but it was a right of freeborn Englishmen. They declared that these disputes should be decided jointly by both employers and employees. However, they sought to replace the judgment of two viewers, who inevitably were perceived as owners' men, with a joint committee of hewers and viewers. "We admire the wisdom of our ancestors," they wrote,

> who appointed that every Englishman should be tried by his peers, a mode the most equitable perhaps that could be imagined. Agreeably thereto, we propose that, for the purpose in question, two Viewers be appointed by the proprietors, and two hewers by the workmen, who shall be empowered to settle all disputes referred to them.[39]

The bond, therefore, could curtail the despotic powers of management. Yet while it is true that miners resisted some changes to the bond that they felt were especially favorable to them – such as the changing of binding time

---

A nearly identical clause appears in the Monkwearmouth Colliery bonds of 1841 reprinted in *Reports from Commissioners: Children's Employment (Mines)*, P.P. (1842), vol. 16, 538.

[35] Flinn, *History of the British Coal Industry*, 65.
[36] *Newcastle Chronicle*, 10 March 1832.
[37] Ibid., 12 May 1832.
[38] See *Newcastle Chronicle*, 5 May 1832, for bargaining at Hetton and 26 May 1832 for bargaining at Burdon Main Colliery.
[39] *A Candid Appeal*, 16.

or the reduction of binding money – by the second quarter of the nineteenth century, it certainly was not true, as Flinn asserted, that the miners generally favored the bond system *tout court*.[40] Nor is it precisely accurate to claim, as Colls has recently argued, that the bond evolved from a "an individual, negotiated contract" into a document of exploitation and control during the first half of the nineteenth century.[41] The negotiated and contractual elements of the bond were always in evidence during this period. Moreover, the bond had always functioned, and continued to function, as one of the basic elements in the owners' control of the labor market.

The miners' unions of the first half of the nineteenth century certainly were premised on the recognition that annual bonds in and of themselves gave undue power and authority to owners and viewers. Despite the limits on managerial authority, the unions unequivocally claimed that the bonds were "an instrument of oppression and fraud."[42] However, they argued this not because the bonds embodied the principle of exploitation through the control of the labor process, but because the institution of the bond undermined workers' power in the labor market. The miners were forced by circumstance either to accept the terms of their masters or to give up their homes and livelihood. The bond, therefore, functioned to more fundamentally structure market relations rather than production relations. The miners demanded a system of industrial relations that not only would embody equity and justice, but also would secure for them significantly increased bargaining power both in the labor market and in the resolution of local colliery disputes.

Notwithstanding the elaboration of the skills and autonomy of the miner that has so intrigued modern labor historians, the social position of miners in the nineteenth century was more akin to that of the common laborer. Adam Smith accounted for the relatively high wage rates of miners with reference solely to "the hardship, disagreeableness, and dirtiness of his work." The tasks themselves, however, were only "the most common labour" situated "in a trade which has no exclusive privilege."[43] A north country doctor, when describing the moral habits of the pitmen to a mines commissioner, noted as an aside that was meant to explain much that "they are, it must be remembered, recruited from the dross of society."[44]

The trade's lack of apprenticeship regulations or other forms of entry

---

[40] Flinn, *History of the British Coal Industry*, 357.
[41] Colls, *Pitmen of the Northern Coalfield*, 64–73.
[42] *A Candid Appeal*, 16. Here the union cited the legal decision of Lord Thurlow: "If the person did not understand the bargain he made, or was so oppressed that he was glad to make it, knowing its inadequacy, it will shew a command over him, which may amount to fraud." See also, Colls, *Pitmen of the Northern Coalfield*, 65; Challinor and Ripley, *The Miners' Association: A Trade Union in the Age of the Chartists* (Lawrence & Wishart, 1968), 94–110.
[43] Smith, *Wealth of Nations*, 116–7.
[44] Leifchild, *Our Coal and Our Coal-Pits*, 218.

restriction was compounded by the fact that miners did not control the entirety of the workplace, only their stall down the pit. The provision of work was generally the prerogative of management, who opened or closed the mines. By the 1830s, it was not unknown for a mine to be open for work only five or six days each fortnight in order to support the cartel's restriction of production. In its early stages, Tommy Hepburn's union of 1831–2 was motivated principally by the desire to force the viewers to provide more work for the miners.[45] Therefore, while it is true that the miners' autonomy was uncontested at the point of production, this should not be construed to imply that the miners controlled the workplace.

The contradiction inherent in defining the miner both as a skilled and autonomous worker and as a common laborer reflected their ambiguous position as both an employee of a large, capital-intensive firm and as a commodity producer. Certainly, the perception that the nineteenth century witnessed the progressive devaluation of the pitmen's craft to that of an ordinary wage laborer misses the point.[46] The problem for the miner was not one of progressive proletarianization under industrial capitalism but one of intrinsic incongruity. The miner certainly was "common" in that he possessed a skill that was reproducible without significant training or apprenticeship and that he depended upon a workplace (but not tools or skills) provided by capitalists. However, the miner also produced a commodity and secured his earnings through payments based upon the quality and volume of the finished product. This entailed a substantial degree of independence and a dependence upon market relations.[47] This profound contradiction between the miner as skilled artisan and as common, alienated laborer lay at the heart of the miners' perception of their work and community and also marked their trade union movements of the early nineteenth century. It not only created the basis of their vision of a system of industrial relations based upon equal and reciprocal power over capital and labor, a vision that grew out of their experience of bargaining at the point of production, but also formed the foundation for the miners' acceptance of collective bargaining practices and then the sliding scale in the second half of the nineteenth century.

Bargaining and market relations

Piece work was the basis of miners' earnings throughout the nineteenth century, and piece rates were subject to revision through bargains struck

---

[45] See Chapter 7.
[46] Colls, *Pitmen of the Northern Coalfield*, 6–44.
[47] A similar experience is noted for French mule spinners by William Reddy, *The Rise of Market Culture*, 161. However, his conclusions differ markedly from those expressed here.

both individually and collectively at the point of production. "The pitman works by bargain altogether," Tommy Hepburn's union proclaimed in 1831, "he works at so much per score, or so much per yard – he gets not a penny, but what is got at the greatest extremity."[48] The nearly endless variations that affected piece rates, including the width of the seam, wet or dry workings, the use of Davy lamps, narrow work measured by the yard, working in the broken, the purity or foulness of the seam, and the like, made it necessary to construct separate piece rates for each pit at each colliery or, as we have seen in the case of Hetton Colliery in the 1830s, separate piece rates for different seams of coal within the same pit.[49]

The historical record is replete with evidence of bargaining at the point of production, but it is evidence that has largely been ignored because it exists beneath the history of the formal institutions and goals of district-wide unionization. The deep game of collective bargaining entailed both subtle and obvious devices employed by both capital and labor. Wildcat strikes at individual collieries were one such tactic, and it is likely that they were frequent occurrences in the early nineteenth century. For instance, T. J. Taylor, the owner of Earsdon Colliery, seemed undisturbed by the work stoppage being undertaken by the hewers when the children's employment commissioners interviewed him in the early 1840s. He matter-of-factly explained that the system of fining hewers for poor work whereby they forfeited, or "laid out," payment for corves sent to the bank with too much small coal "caused a good deal of grumbling amongst the men, and they occasionally, when there is a great number of corves laid out, lay the pit idle, as has happened this day."[50]

Perhaps a more delicate maneuver in the collective bargaining game included the suggestion of competition for labor. In 1804, for example, one hewer at Hartley Colliery spread the news that an overman from neighboring Cowpen Colliery had offered him over ten guineas to bind at that colliery, an offer that later was disavowed by the Cowpen owners. However, as the Hartley Colliery manager saw it, the rumor "may be to mislead, but we shall have the best information we can obtain before any thing is gone rashly into or believed, where it is so much [in] the Workmens Interest to mislead."[51]

Whereas workers might hope to bring pressure on their employers to raise rates by circulating rumors of higher earnings elsewhere, owners hoped to avoid labor struggles by first winning over the local leaders at their colliery.

[48] N.E.I.M.M.E., Bell Collection, vol. 11, 284, "To the Public: The Pitmen of the Tyne and Wear."

[49] See Chapter 3.

[50] *Reports from Commissioners: Children's Employment (Mines)*, P.P. (1842), vol. 16, 608.

[51] N.R.O., Delaval Papers, ZDE 4/26/64, Bryers to Delaval, 3 October 1804; ZDE 4/26/65, Bryers to Delaval, 7 October 1804.

Thus, Jonathan Bryers at Lord Delaval's Hartley Colliery withheld making an offer of piece rates to his workmen until "we shall be able to judge what influence we can have with the principals."[52] Bryers, for one, certainly worked hard at winning them over. A week later, he wrote to Lord Delaval that "during the week & especially on Friday Evening last, we endeavoured all we could to bring some of the leading men over but without *full* effect."[53]

For Tyneside hewers in the early nineteenth century, Newcastle became the center for the exchange of information concerning piece rates, fines, and binding money offered in the region. In 1804, miners connected with a somewhat inchoate union movement sometimes called the Brothered Men "were determined not to bind until they saw what was doing at N[ew]Castle ... and the offers made by the respective collieries."[54] At binding times, Newcastle took on the aspect of a hiring fair. Viewers and agents, however, were adamant that bindings should not take place there because it might lead to competitive bidding for labor. When Lord Delaval suggested to his agent that extra hewers could be hired in Newcastle, he was advised:

> It is a fact that if we had begun to bind 20, 40 or 60 men there, before five could have been agreed with, the whole of the Agents would have been authorized by such example to do the best for their respective Collieries and ... it would be a most impolitic step ... in opposition to the whole Coal Trade & so much against the interest of the whole, both in Money & work.[55]

The competitive market that might develop for labor under these conditions was so feared that during the 1805 binding, the coal owners agreed among themselves not to allow any agents or viewers "to be in or near Newcastle ... so that the men might not be encouraged to expect any other offers to be made to them etc."[56]

These tactics, nevertheless, were only part of the broad mosaic of collective bargaining that revolved about the site of production in the early nineteenth century. To some extent, as J. W. F. Rowe noted for the later nineteenth century, this process was a hidden one often because it was continuous. Since the productive process underground changed both constantly and irregularly as coal seams and working conditions were altered, price lists were continually revised to take into account new hewing conditions.[57] Thus, there was never any such thing, even in the era of district-wide bargaining and the sliding scale of the late nineteenth century, as a "standard" rate or

---

[52] N.R.O., Delaval Papers, ZDE 4/26/64, Bryers to Delaval, 3 October 1804.
[53] Ibid., ZDE 4/26/66, Bryers to Delaval, 14 October 1804.
[54] Ibid.
[55] Ibid.
[56] Ibid., ZDE 4/27/88, Bryers to Delaval, 20 October 1805.
[57] Rowe, *Wages in the Coal Industry*, 49–51, 120.

"fixed" price list. The very ubiquity of change and bargaining perhaps has even led to its being overlooked by historians or, in one case, to its being dismissed as anachronistic.[58] However, the reconstruction of bargaining practices in the early nineteenth century through a variety of legal sources, published accounts, and private correspondence leaves little doubt that bargaining at the point of production was the principal, although not always ultimate, forum for industrial relations.

Already in 1805, the manager of Hartley Colliery referred to a pitmen's petition as "according to custom."[59] The pitmen's demands at that time were quite extensive and reflected their favored position in the labor market during the Napoleonic Wars. The Hartley colliers demanded increased piece rates per score of both large round coals sent to the bank and small coal; the provision of wheat and rye at subsidized prices – a practice that was probably discontinued when grain prices fell;[60] the provision of powder for blasting and candles for work underground; and, binding money, or a signing bonus, of fourteen guineas.[61]

In that year, the men resisted signing the bonds for nearly a fortnight, and even though the Hartley agent expressed a slight fear that the men might begin to leave the colliery to seek employment elsewhere,[62] the owners effectively united to limit the movement of labor. Eventually, the men at Hartley gave up their petition after hearing that some men at neighboring Cowpen Colliery had been refused employment there and were on their way to Hartley. This, according to Bryers, "set them a going & very little difficulty was afterwards experienced, except in keeping back those that we supposed were scarely able to work a reasonable day's work."[63] Yet negotiations over the bond continued up to the very day of the signing. "The men," it was noted, "kept up their good understanding amongst themselves to the very last, & had several deputations at us on Friday still lowering their demands, which was repeatedly resisted."[64]

At times, the breakdown of collective bargaining could lead to popular action, especially if blacklegs were brought in to replace the recalcitrant workers. For example, the violence that attended the riot at Waldridge Colliery on Christmas Eve, 1831, the so-called Waldridge Outrage, was in fact the result of a long-simmering dispute between the miners and their

[58] Colls, *Pitmen of the Northern Coalfield*, 73.
[59] N.R.O., Delaval Papers, ZDE 4/27/86, Bryers to Delaval, 13 October 1805.
[60] Flinn, *History of the British Coal Industry*, 381–2.
[61] N.R.O., Delaval Papers, ZDE 4/27/86, Bryers to Delaval, 13 October 1805.
[62] Ibid., ZDE 4/27/88, Bryers to Delaval, 20 October 1805.
[63] Ibid., ZDE 4/27/89, Bryers to Delaval, 27 October 1805.
[64] Ibid.

employers that had occurred after the collapse of informal collective bargaining.

The viewer, Anthony Seymour, testified at the trial of those indicted for rioting that the colliery had been working only since 1830 and that disputes over "workings and wages" had forced the owners to employ blacklegs. Mr. Cookson, the prosecutor, explained in his opening remarks that the dispute could be traced back to the opening of a new seam of coal in the mine and the dissatisfaction expressed by the men with the piece-rate prices offered for working this seam. Upon cross-examination, the defense attorney, Mr. Archbold, got the viewer to admit that management had exhibited a significant degree of confusion over exactly how to settle the dispute. The viewer first suggested inviting an arbitration team of sixteen men comprised of eight pairs of viewers – one pair each selected by the men and the owners – from each of the surrounding collieries to settle the matter.

This plan was soon abandoned, and instead, a team of two arbitrators, Thomas Seymour for the masters and W. Longstaff for the men, met and fixed a list of prices that the owners accepted but that were subsequently rejected by the men. At that time, about fifty miners left off work at Waldridge and were replaced by lead miners recruited from west Durham. Seymour, the Waldridge viewer, appeared testy under cross-examination and reiterated his dedication to the arbitration system: "Two viewers were appointed," he was reported to have said, "one by the men and one by the masters, to arbitrate between them. The arbitrators did not differ. He [Seymour] still has the decision of the viewers in his pocket."[65]

The hidden history of collective bargaining is similarly revealed in the case of Burdon Main Colliery in 1831 and 1832. The miners at that colliery had accepted employment and begun work before Tommy Hepburn's union had firmly coalesced. Within days of accepting and signing their bonds at the old piece rates, they quit work and joined Hepburn's union movement. When the union succeeded in gaining significant wage advances in 1831, the owners of Burdon Main, "with a liberality which ought to have produced in the Workmen better feelings and conduct," granted piece-rate increases to hewers equivalent to an advance of about 10 percent and those to putters equal to 9 percent, as well as an allowance of two pounds of gunpowder each fortnight to hewers for work in the Low Main Seam.[66]

In the following year, 1832, a fortnight before the commencement of binding time, the miners presented a petition to their viewer, Mr. Johnson, for further adjustments of their piece rates. The petition demanded an in-

[65] *Newcastle Chronicle*, 10 March 1832.
[66] N.E.I.M.M.E., Bell Collection, vol. 11, 417.

crease of sixpence per score, about 9 percent, on work in the Low Main Seam where it was five feet wide. (The previous year's increase had been identical, sixpence per score, but applied only to areas where the seam was four feet nine inches wide.) Furthermore, the petition requested an increase of sixpence per yard for yard work such as winning headways and holing walls.[67]

In response, the viewer rejected any increase of rates per score, but offered an advance of twopence per yard for yard work. In the course of "various meetings [that] took place between Mr. Johnson and the Men," the miners relinquished their demand for increases in piece rates per score but, it seems, insisted on the sixpence increase in yard work prices. By binding day, 5 April, the dispute still had not been settled, and the viewer expressed a certain willingness to hire all workers nonetheless, except six men who were undoubtedly targeted as union activists.[68]

The miners of Burdon Main refused to acquiesce in the victimization of their organizers and struck the colliery once again. After three weeks, the miners presented a second petition to the viewer, this time insisting upon the rehiring of all men bound the previous year. They also tweaked the noses of the owners by demanding an increase in the day rates of "shifters," the superannuated hewers who worked aboveground. This new demand had been inspired by the recent publication of a report by the coal owners' cartel claiming that shifters' wages were regularly three shillings per day throughout the northern coalfield. At Burdon Main, they were paid between 2s. 6d. and 2s. 10d. The increase in shifters' rates was necessary, the men claimed, "in order . . . that you might not falsify the statement of your worthy Chairman."[69]

The conclusion of the dispute at Burdon Main Colliery has not survived, but the very disappearance of evidence suggests that the parties reached some form of accommodation and consensus. The ubiquity of local collective bargaining and arbitration procedures certainly runs counter to the cataclysmic picture of early-nineteenth-century industrial relations as well as to Robert Moore's contention that the northern miners' accommodationist attitudes were the result of Methodist ideology.[70] On the one hand, this is partly because the existence of the bond became the fulcrum of district-

[67] Ibid.; *Newcastle Chronicle,* 26 May 1832, "The Late Pitmen of Burdon Main Colliery to their Owners."

[68] N.E.I.M.M.E., Bell Collection, vol. 11, 417.

[69] *Newcastle Chronicle,* 26 May 1832.

[70] Flinn's account of industrial relations before 1830, for example, virtually ignores any distinction between district-wide and local organization. See Flinn, *History of the British Coal Industry,* 367–411. Colls's concentration solely on the disputes over the terms of the bond led him to claim mistakenly that by 1832 bargaining was an anachronism in the northern coalfield. See Colls, *Pitmen of the Northern Coalfield,* 73. Moore's study is *Pitmen, Preachers and Politics.*

wide organization and thus has received greater attention from historians. Industrial relations at the level of the workplace have consequently been ignored or submerged beneath these more widely publicized concerns. Moore's contention, on the other hand, lacks an adequate understanding of the history of industrial relations in the northern coalfield and thus posits a role for religious ideology that justly belongs to the organization and structure of industrial relations. Contrary to these interpretations, workplace bargaining was and continued to comprise the substance of industrial relations throughout the nineteenth century.

The recognized system of workplace bargaining in the northern coal industry accorded a degree of market power to workers that required a vigilant defense. Coal owners certainly would have preferred to establish uniform piece rates throughout the entire region and to do away with bargaining altogether. This would have had the effect not only of subverting the workers' power to bargain in particular, but of undermining the labor market generally. The establishment of uniform piece rates, for example, was attempted after the union movement of 1809–10. But by 1812, the attempt to erect "some system of equating the different wages so as to be beneficial to the Trade" had foundered. The coal owners were forced to accept the apologies of their committee who were "sorry to say that from the charges being so variously intermixed they are disappointed in their expectations."[71]

Certainly by the 1820s, the owners were reluctantly forced to accept the role of workplace bargaining, especially with regard to hewers' piece rates. Thus, the limited attempt of the Lambton and Londonderry collieries to regulate wages in 1822 was based on the understanding that "we have not however ventured to meddle with the *Coal Hewers*."[72] Less than a decade later, in 1831, the coal owners' first published response to the demands put forth by Tommy Hepburn's union admitted that "the Coal Owners conceive that their men are paid higher than any other of the same description in the kingdom," but they also denied acting as a group to influence piece rates. "With the rate of wages at the respective collieries," the coal owners' cartel explained, "the Coal Owners do not, as a body, interfere, the prices being entirely governed by local circumstances, and left to the discretion of the Viewers."[73]

The significant market power embodied in workplace bargaining over piece rates was protected to some extent by its separation from conflicts and

[71] N.R.O., Committee Meetings of the Coal Owners of the Tyne and Wear, 12 March 1812, also quoted in Colls, *Pitmen of the Northern Coalfield*, 83.
[72] D.C.R.O., Londonderry Papers, D/Lo/C 142, Buddle to Londonderry, 24 March 1822.
[73] N.E.I.M.M.E., Bell Collection, vol. 11, 368.

negotiation over the terms of the annual bond. The United Colliers of 1825–6, for example, published a copy of a model miners' bond. They reprinted the following final clause to which was appended the union's own explication of the independence of piece-rate bargaining from the bond (original emphasis):

> *And, Lastly,* That the rates and prices to be paid by the said A. B. their executors, administrators, or assigns, to the said several other parties, of the one part, shall be as follows; viz. –
> [*Here follow the rates and prices, which are regulated according to the local circumstances of each Colliery.*][74]

Similarly, John Buddle recognized that Tommy Hepburn's union was organized on the basis of a regional body that sought the redress of inequities in the bond while local bodies bargained for changes in piece rates, fines, workmen's charges, and the measurement of production. He wrote to Lord Londonderry in April 1831:

> I have made out, pretty clearly, the plan of their *Union* – it is in two Branches, General and Local. The general union extends only to 3 points viz. 11 days in the fortnight, 12 Hours for the Boys, and not to be turned out of their houses....
>
> The local Union goes to the measure of the Corves, the finding [provision] of Candles, Gunpowder, etc. and to certain items of the Work in individual Collieries.[75]

Even after the 1877 adoption of the sliding scale, which was supposed to have pegged wages to the price of coal in Durham, and the later acceptance of district-wide conciliation boards in the late 1880s and early 1890s, bargaining at the point of production was perhaps the most crucial aspect of the determination of wage rates and earnings at individual pits and collieries.[76] J. W. F. Rowe's study of wages in the coal industry between 1888 and 1922 warned that changes in piece rates took place at individual collieries "irrespective of the recognized district percentage variations" because price lists "cannot in the nature of things be fixed once and for all, or be universally applicable."[77] Rowe stipulated that despite the existence of formal institutions of collective bargaining and the linkage of piece rates to prices, in the short-term, wages were more responsive to the "hidden alterations" that took place

---

[74] *A Candid Appeal,* 9.
[75] D.C.R.O., Londonderry Papers, D/Lo/C 142(675), Buddle to Londonderry, 13 April 1831.
[76] On the legislative developments of midcentury and the development of the sliding scale, see Neil K. Buxton, *The Economic Development of the British Coal Industry* (London: Batsford, 1978), 135–9; 150–6; Church, *History of the British Coal Industry,* 697–702.
[77] Rowe, *Wages in the Coal Industry,* 49.

at the point of production and that were "directly influenced by the bargaining power of labour."[78]

Moreover, Rowe expressed two fundamental caveats to his study of wages in the coal industry: first, that on-site bargaining regularly altered piece rates; and, second, that collective bargaining agreements at the district level therefore were "not necessarily an accurate measurement of the wage levels" of labor in the mines. He concluded:

> The alteration of price lists is simply and solely a matter of bargaining.... The employer will maintain that the hewers should be able to send out X tons a day without undue effort; the hewers will put it at x/2; both parties would nominally agree that their object is to fix such a price per ton as would give roughly the normal earnings current in the district, but actually each is out to get an advantage, and the final result will depend on bargaining power."[79]

Naturally, this is not to say that workers in the early nineteenth century defended workplace bargaining because they successfully brought their market power to bear on every case. As we have seen, the viewers' tactics and understanding of the game was on a par with the miners, and their resources were much greater. However, workplace bargaining was recognized by the miners as a source of independence and market power. Interestingly enough, James Losh, a northern coal owner and attorney, believed that this adversarial market relationship lay at the heart of the unstable social relations in the northern coal industry. He noted in his diary that industrial relations could be conducted peacefully if "the coal owners [were] to act *honestly* and *cordially* together." He then went on to suggest:

> We [the coal owners] should act openly and fairly in every thing which relates to wages, etc., whereas I fear that the Viewers have not only been harsh in their manner towards the men but have also endeavoured to *get the better* of them, as it is called, in bargains, the mode of working, etc.[80]

The dominant role that piece-rate bargaining played in the working lives of miners throughout the nineteenth century further functioned to reproduce the contradictions inherent in piece work, a contradiction described by Marx. One of Marx's more widely known claims is that piece work was "the form of wage most appropriate to the capitalist mode of production" because it forced laborers not only to supervise their own product but also to exploit themselves by working longer hours to gain greater wages. Marx noted, however, that piece work also tended to foster "the workers' sense of liberty,

[78] Ibid., 120.
[79] Ibid., 50–1.
[80] Hughes, *Diaries of James Losh*, 2: 113.

independence, and self-control" by creating the illusion that they actually controlled and sold the product of their individual labor.[81]

This duality of piece work was a reflection of the ambiguous social position of the northern miners. On the one hand, miners were common laborers – "the lowest of the low, (in common estimation)," the pitmen's union of 1831 wrote[82] – disposing of their alienated labor, degraded by piece work that encouraged self-exploitation, and creating a culture that was seemingly perverse in the value and pride taken in Stakhanovite efforts.[83] On the other hand, the miners defended their skill, expertise, and independence in the workplace. In the 1820s and 1830s, for example, when the coal owners tried to compare the relatively high piece rates of mining to the lower wage rates of other workers, such as southern agricultural laborers, the miners invariably invoked the skills and dangers of coal mining in their own defense. Thus Tommy Hepburn's union claimed that "if the coal hewer laboured like other labourers his earnings would be like the earnings of labourers in the South, or worse. [But] the sun looks not upon any kind of labour that can in the least be compared to hewing and putting."[84] In the same vein, "an old pitman" who defended the union movement of the 1830s argued that a hewer's skills were developed over a long period of time and that the "habits of industry necessary for working a pit are not easily acquired."[85]

Finally, the acceptance of piece-rate bargaining as a symbol of the skill and independence of the hewer entailed an important degree of acceptance of a market ideology, that is, the acceptance in some form of the efficacy of supply and demand. There was, not surprisingly, some confusion expressed concerning which market mechanisms prevailed over their circumstances: those of the labor market as befitted laborers or those of the commodity market as befitted independent producers. In the first instance, the miners' protest against the annual bond was frequently formulated as a demand to subject the labor market to the forces of supply and demand and thereby weaken the market power of the coal owners. For example, the 1825 United Colliers argued that the bonds were all alike, binding time was the same

---

[81] Marx, *Capital*, vol. 1, 694–8.
[82] N.E.I.M.M.E., Bell Collection, vol. 11, 284, "To the Public. The Pitmen of the Tyne & Wear in Answer to the Tyne Mercury and other Publications," 18 April 1831.
[83] "Hewing matches," for example, were a part of the pit culture of the early-nineteenth-century northern coalfield; see Robert Colls, *The Collier's Rant* (London: Croom Helm, 1977), 65; and idem, *Pitmen of the Northern Coalfield*, 32–3. The term "Stakhanovite" is not as anachronistic as it may first appear. Stakhanovism was in fact a part of the restructuring of Soviet industrial relations that reemphasized piece work. See Moshe Lewin, "Social Relations inside Industry during the Prewar Five-Year Plans, 1928–41," in idem, *The Making of the Soviet System* (New York: Pantheon, 1985), 253–4.
[84] N.E.I.M.M.E., Bell Collection, vol. 11, 284.
[85] *Newcastle Chronicle*, 17 March 1832.

throughout the coalfield, and the coal owners conspired so that "the poor collier shall have no choice, no alternative."[86] Ben Embleton, a union leader both under Hepburn in the 1830s and with Martin Jude in the succeeding decade, explained in 1843 that "the bond was concocted in the coal trade office and the coalowners took good care to have the binding all their own way."[87] In the midst of the 1831–2 miners' strike, Tommy Hepburn simply explained that his union "wanted an agreement, but, at the same time, they wanted one-half of that agreement to be of their own making."[88]

In the second instance, when the miners acted as commodity producers, they often expressed a similar willingness to accept the dictates of the product markets. Employing a logic similar to that used in the analysis of the labor market, free trade in commodities was meant to break the market control of the coal owners. Hepburn's union published a handbill in April 1831 proclaiming "that the Men and Boys are willing to abide the Risk of the Fluctuations of the Coal Trade."[89] However, here again the miners' unions had to fight against the owners' subversion of market forces. The coal owners' cartel, they argued, kept market prices high by limiting the supply of coal. This reduced the amount of work available to miners. Correspondingly, wage rates might be high, but earnings were actually quite low. "The reason why the owners give no better work," another union handbill announced, "is of course, from the combination among themselves, to send only a certain quantity of coals to market, in order to keep up prices."[90]

Therefore, the acceptance of some form of market ideology long predated the classic period of labor's accommodation to capitalism after 1850. The sliding scale and the district conciliation boards of the later nineteenth century were built upon these structures of work and industrial relations that were perceived as and acted upon as market relations. There were profound contradictions inherent in the social position of miners, but these did not vitiate the miners' general adherence to market principles, particularly as it was evident in their defense of workplace bargaining.

The miners understood social relations with the coal owners largely in terms of the market. Whether expressing their opposition to the coal owners in a language appropriate to laborers or commodity producers, the early miners' unions identified the control of the market, and not the labor process, as the source of class power. The defense of workplace bargaining was one of the principal sources of market control, for it entailed the mutual accep-

---

[86] *A Candid Appeal*, 16.
[87] *Gateshead Observer*, 18 March 1843, quoted in Challinor and Ripley, *Miners' Association*, 94.
[88] *Newcastle Chronicle*, 23 June 1832.
[89] N.E.I.M.M.E., Bell Collection, vol. 11, 226.
[90] Ibid., 217.

tance of the reciprocal power of capital and labor in the workplace. However, the experience of unionization under Tommy Hepburn in 1831–2 would reveal that securing workplace bargaining was not enough to defend labor's position within the system of industrial relations. As a result, union tactics, it will be argued in later chapters, evolved to include the formal control of the movement of labor as well as the control of output. The combination of these tactics eventually secured for Hepburn's union control of both the product and labor markets. This marked a radical reconstruction of market relations as well as a temporary redistribution of class power.

Bargaining, custom, and pit culture

The prevalence of bargaining in labor relations could not help but find some reflection in the ethics and values of the northern miners. Bargaining, for example, acted to reinforce the manly ethic of the pit. As already described, John Buddle, who was harassed unceasingly throughout the 1831 miners' strike, was likewise feted at the conclusion of the strike by his employees for fighting them fairly and like a man.[91] Pit bargaining, as Bill Williamson has written, was an essential part of the skill and expertise of the hewer.[92] Thus, to be a man down the pits was to be a strong worker and a safe worker, as well as a tough bargainer with management. "Rate busters," like those used at Esh Winning Colliery in the 1920s to drive down piece rates, were likely to be deemed stupid or viewed as the manager's lackeys.[93]

The need to be a skilled and tough bargainer with a keen head for figures may also have led miners to direct their own education toward mathematics. Children's Employment Commissioner J. R. Leifchild recognized that a "reading pitman will have no political books; but one or more mathematical works, picked up from a stall at Newcastle."[94] Similarly, Robert Lowery, the North Shields radical who supported the miners in the 1820s and 1830s, recorded in his autobiography that among the miners, "there were many superior mathematicians, and the booksellers of Newcastle were known to sell, chiefly among the workmen of the north, a larger number of works on that science than were sold in any other similar district of the country."[95]

While it has been argued here that bargaining and market relations played a preeminent role in the working lives of northern miners in the first half of the nineteenth century, this is not to imply that custom played absolutely

---

[91] See Chapter 3.
[92] Williamson, *Class, Culture and Community*, 93.
[93] Moore, *Pit-men, Preachers and Politics*, 86.
[94] Leifchild, *Our Coal and Our Coal-pits*, 218.
[95] Brian Harrison and Patricia Hollis (eds.), *Robert Lowery: Radical and Chartist* (London: Europa, 1979), 78.

no role. However, drawing the distinction between "customary" and "modern" work practices has increasingly become a hazardous enterprise. As Michael Sonenscher has shown in the case of eighteenth-century French hatters, customs were the creation of the vibrant and active assertion of artisanal rights that in and of themselves were not immune to market pressures.[96]

However, if for the moment we accept the dichotomy between customary and market attitudes toward work, then drawing cavils may be considered a customary and nonmarket approach. To some extent, also, a customary attitude toward work was even embodied in the bonds that required the hewers to "do and perform a full day's work on each and every working day, or such quantity as shall be fairly deemed equal to a day's work."[97] A fair day's work was commonly considered by both miners and owners to be equal to an eight-hour shift, a limitation that was eventually incorporated into the bonds by the 1840s.[98]

A full or fair day's work as measured in hours, however, might allow very good hewers who happened to draw very easy cavils to earn extraordinary wages while a weaker worker laboring under more difficult circumstances might earn very little. This unfairness of chance and circumstance was likely to have contributed to the development of output restriction as measured by limitations on daily earnings as a form of work regulation. Carbonarius, a supporter of Hepburn's union, was very careful to point out that only the best workers and those who worked in the most dangerous parts of the mine could hope to earn a fair day's wage, in this case four shillings, for a fair day's work. By the 1830s, these customary earning limitations may actually have been more of an ideal than a practical reality, but they acted as visible evidence of workers' control underground.[99]

---

[96] Michael Sonenscher, *The Hatters of Eighteenth-Century France* (Berkeley and Los Angeles: University of California Press, 1987), xi, 18–31. I would like to thank Isser Woloch for directing me to this work. See also E. Hobsbawm and T. Ranger (eds.), *The Invention of Tradition* (Cambridge University Press, 1983); G. Stedman Jones, "Class Expression versus Social Control?: A Critique of Recent Trends in the Social History of 'Leisure,' " reprinted in idem, *Languages of Class*, 84–5; and Neil MacMaster, "The Battle for Mousehold Heath 1857–1884: 'Popular Politics' and the Victorian Public Park," *Past & Present*, no. 127 (1990), 136–8.

[97] *Reports from Commissioners: Children's Employment (Mines)*, P.P. (1842), vol. 16, 537. The exact phrase appears in the 1826 bonds; see *A Candid Appeal*, 8.

[98] *Rules and Regulations for the Formation of a Society, to be Called The United Association of Colliers on the Rivers Tyne and Wear* (Newcastle-upon-Tyne, 1825), 6; "Report by the Committee of the Coalowners Respecting the Present Situation of the Trade," *Newcastle Chronicle*, 17 March 1832; *Reports from Commissioners: Children's Employment (Mines)*, P.P. (1842), vol 16, 537.

[99] *Newcastle Chronicle*, 24 March 1832; James A. Jaffe, "The State, Capital, and Workers'

Output and earning restrictions served other functions as well. Robert Colls has argued that restriction was above all a union policy designed to curb the hewer's autonomy. According to Colls, the autonomy and independence of the miners served to undermine their collective interest by encouraging competition between hewers and "an anarchy of production."[100] Evidence presented below, however, suggests that at least in 1831–2, restriction was of rank-and-file origin and was accepted only reluctantly by the trade union leadership. Thus, the degree to which restriction embodied a cultural assault on customary attitudes toward work remains an open question.

Several contemporaries, moreover, believed that restriction constituted an attempt to create and distribute more jobs, an interpretation that might account for its genesis as a rank-and-file demand.[101] John Reay, the cashier at Wallsend Colliery, wrote to John Buddle in 1832 that "at several of the Collieries they [the hewers] will not work for more than 3/. per day – 'in order to get All the Men Bound.' "[102] Buddle himself told Lord Londonderry that "your Lordship's Men are anxious to work a number of days for the sake of getting the Boys and old men (Shifters) large wages, but the Hewers will not work more than a limited quantity."[103] Henry Morton came to the same conclusion:

> One of the articles of the union [in 1831] enjoins that the men employed shall not earn above a certain turn a day, perhaps 3^s-[,] that employment may be given to a greater number of hands. If this conjecture prove[s] correct we shall bind a large number of course – The Pitmen are a most singular race, and are really possessed of all the Crafty wiles of the American Indians.[104]

Restriction, therefore, could serve two functions: First, it could create a demand for more hewers in order to maintain a colliery's output. Second, in an era when collieries might work only eight or nine days a fortnight to support the Vend's prices, restriction could also extend the number of days a colliery was forced to work in order to maintain that level of production. In this manner, both piece-rate workers and day workers supported restriction. John Hall, a leader of the Durham and Northumberland miners during

Control during the Industrial Revolution: The Rise and Fall of the North-east Pitmen's Union, 1831–2," *Journal of Social History*, 21, no. 4 (1988), 722–3.

[100] Colls, *Pitmen of the Northern Coalfield*, 29–33.
[101] On this point, see Carter Goodrich, *The Frontier of Control*, 178. Richard Whipp also makes the point that among pottery workers, "stints" were regularly adopted in order to equalize the distribution of work. See, " 'A Time to Every Purpose': An Essay on Time and Work," in Joyce, *Historical Meanings of Work*, 227.
[102] D.C.R.O., John Buddle Papers, NCB I/JB/1188, Reay to Buddle, 2 April 1832.
[103] D.C.R.O., Londonderry Papers, D/Lo/C 142(715), Buddle to Londonderry, 29 May 1831.
[104] Lambton Mss., Morton to Durham, 17 May 1831.

the 1840s, calculated that a reduction in output of about 25 percent by each hewer could increase the number of working days by about fifty each year, resulting in about an £8,000 increase in the earnings of day workers throughout the northern coalfield.[105] Therefore, as the evidence indicates, output restrictions altered the terms of market relations. The labor market came under the functional influence of the workers by creating an increased demand for labor. Furthermore, as we shall see, output restrictions naturally served to support commodity prices, the informal index upon which piece-rate prices were often pegged.

Practices such as cavilling and restriction, therefore, may have in part represented customary, nonmarket attitudes toward work in that they can be considered as attempts to deal with the inequities of employment that easily could have been turned against the miners. Nevertheless, they were imbedded in a culture of work that accepted the market as the proper terrain for labor relations. Miners' unions of the early nineteenth century struggled for power in the market, not its abolition. And, given the social and political world of early industrial capitalism, this eventually posed as great a threat to the established order as overtly political working-class movements.

[105] *Miners' Advocate*, 16 December 1843, quoted in Challinor and Ripley, *Miners' Association*, 111–12.

6

Religion, ideology, and trade unions

The influence of evangelical religion upon the British working class revolves about one of the implicit themes of this work: the accommodation of the working class to early industrial capitalism. It has already been argued that while industrial relations were inevitably contentious in the first half of the nineteenth century, they were nonetheless anchored in the mutual acceptance of the ideology of the marketplace principally through the practice of workplace bargaining over piece rates. Moreover, as will be shown in succeeding chapters, labor struggles in the early-nineteenth-century coal industry reflected the contest for control of the labor and product markets rather than for control of the labor process.

This argument runs counter to at least one tradition of historical and sociological work that posits religion as one of the principal forces acting to accommodate or subject the working class to the new industrial order. Elie Halévy is perhaps the name most often connected with this thesis.[1] However, it is hoped that by directing attention toward the fundamental importance of the role of the workplace and industrial relations in the creation of ideology, the role of evangelical religion in the lives of the northern working class can be placed in a different perspective. Rather than acting as an agent of submission or accommodation, evangelicalism during this era was the religion of a distinct minority in the coalfields, and its connection to the early labor movement was both pragmatic and ephemeral.

Max Weber's work has necessarily had a profound impact upon the controversy over religion and the working class during the Industrial Revolution, and it is worthwhile pointing out some of its aspects that have influenced the study of British social history.[2] Weber contended that religions of salvation,

[1] Elie Halévy, *A History of the English People in 1815* (1924; repr. London: Ark, 1987), 370–4. Portions of this chapter have previously appeared in my "The 'Chiliasm of Despair' Reconsidered: Revivalism and Working-Class Agitation in County Durham," *Journal of British Studies*, 28: no. 1 (1989), 23–42.

[2] The relevant historical literature is reviewed by Hugh McLeod in *Religion and the Working Class in Nineteenth-Century Britain* (London: Macmillan Press, 1984). On the place of religion within the broad context of Weberian theory, see Wolfgang Schluchter, *The Rise of Western*

such as Calvinism and Methodism, could significantly influence the conduct of daily life by imposing or evoking a religious "systematization of practical conduct."[3] There is a tendency, Weber claimed, for salvationist religions to evolve rationalized and "methodical procedures of sanctification" that "appear to provide a secure and continuous possession of the distinctive religious acquirement."[4] Weber labeled "ascetic" those personal characteristics that combined this "methodical procedure for achieving religious salvation" with the knowledge that one's behavior was directed by God.[5]

Asceticism could take two forms: "world-rejecting" and "inner worldly." In the former, withdrawal from the temptations and diversions of everyday life was viewed by the ascetic as necessary to gain salvation. In the latter, the ascetic operated "within the institutions of the world but in opposition to them."[6] To the inner-worldly ascetic, life became a ceaseless "challenge for the demonstration of the ascetic temper and for the strongest possible attacks against the world's sins."[7] In order to meet this challenge, the individual required, above all, the internalization of rational and systematized conduct in order to secure the "alert, methodical control of one's own life and behavior."[8]

Weber's claim that religions of salvation rationalized and systematized personal conduct became one of the bases for E. P. Thompson's analysis of Methodism's effects on the English working class. Thompson argued, however, that the evangelical message operated largely as an agent of capitalist domination helping to subordinate the industrial working class to the dominion of factory time and work discipline. He expressed a certain caution over the manner and the degree to which this harsh message was internalized, but suggested, nevertheless, that the English working class accepted Methodism only reluctantly and in the aftermath of actual political defeats that marked their social and economic subordination to capital.[9] This view has certainly gained a wide acceptance among many of the most prominent labor historians, including E. J. Hobsbawm and George Rudé, both of whom have maintained that evangelicalism was, in Thompson's words, the working

Rationalism: Max Weber's Developmental History, Guenther Roth, trans. (Berkeley and Los Angeles: University of California Press, 1981); and Tom Bottomore, *Theories of Modern Capitalism* (London: Allen & Unwin, 1985), 22–34.

[3] Max Weber, *Economy and Society*, 2 vols., Guenther Roth and Claus Wittich (eds.), 4th ed. (1956; repr. Berkeley and Los Angeles: University of California Press, 1978), vol. 1, 528.

[4] Ibid., 538.

[5] Ibid., 541.

[6] Ibid., 542.

[7] Ibid., 543.

[8] Ibid., 544. Of course, the most widely read account of this process is Weber's *The Protestant Ethic and the Spirit of Capitalism*, Talcott Parsons, trans. (New York: Scribners, 1958).

[9] Thompson, *Making of the English Working Class*, 350–400.

class's "chiliasm of despair," which, Rudé later wrote, "offered the one-time labour militant... compensation for temporal defeats."[10]

Weber's influence was even more profound and explicit in Robert Moore's study of the effects of Methodism on a late-nineteenth- and early-twentieth-century Durham mining community. In part, Moore set out to provide a case study of Weber's theory of inner-worldly asceticism, but he also hoped to contribute to the historical debate over the function and meaning of Methodism to the northern working class.[11] His principal claim was that Methodism acted to "inhibit the development of class consciousness and reduce class conflict."[12] Methodism, according to Moore, was essentially an individualistic religion that emphasized personal regeneration and individual ethical behavior.[13] Thus, Methodism and Methodist trade union leaders placed greater emphasis upon the individual and respectability than upon class and solidarity; upon vertical divisions of society according to personal qualities and status rather than on horizontal divisions based on class.[14] In an admittedly vague way, Moore further argued that this perception of the world made Methodism more amenable to liberalism and the ideology of the market because it accorded with the Methodists' desire to "develop their talents and measure their diligence."[15]

Weber's influence upon the historians of religion and the working class has certainly not been absolute. In contrast to these sociohistorical treatments of religion, others have investigated the relationship between religion and the people in terms of what might be labeled the political history of religion in early-industrial Britain. Among the most important of these, W. R. Ward has claimed that Methodism was popular among the laboring classes of the early nineteenth century precisely because it complemented political radicalism. Just as political radicals fought Old Corruption, placemen, and boroughmongers in the state, so Methodists sought to reform the complacency, corruption, and indolence of the official Church.[16] David Hempton has

---

[10] George Rudé, *Ideology and Popular Protest* (New York: Pantheon, 1980), 92. See also E. J. Hobsbawm and George Rudé, *Capital Swing* (1969; rev. ed., Harmondsworth: Penguin Books, 1973), 251. Hobsbawm's view here differs considerably from his earlier views on evangelicalism and radicalism in "Methodism and the Threat of Revolution in Britain," *History Today* (1957) reprinted in idem, *Labouring Men: Studies in the History of Labour* (London: Weidenfeld & Nicolson, 1964).

[11] Moore, *Pit-men, Preachers and Politics*.

[12] Ibid., 26.

[13] Ibid., 26–7, 113–19.

[14] Ibid., 156.

[15] Moore studiously avoids any causal linkage of religion and Liberalism, but posits more vaguely Weber's notion of an "elective affinity" between the two. See Moore, *Pit-men, Preachers and Politics*, 23, 158–9.

[16] W. R. Ward, *Religion and Society in England, 1790–1850* (London: Batsford, 1972); see also

concluded similarly that Methodism's appeal to sections of the British working class in the early nineteenth century was based not on the chiliastic response of defeated workers but on the "missionary optimism" of English Methodism, which in its origins was tolerant, Arminian, individualistic, and democratic.[17]

Finally, A. D. Gilbert's work provides an alternative sociological analysis of the appeal and effect of Methodism. Gilbert placed Methodism within the context of a general evangelical revival in Britain that reflected the growing economic and social dislocation brought on by industrialization. "Evangelical nonconformity," as he termed it, served a number of functions: to emancipate its followers from the paternalism of the preindustrial world; to satisfy the anomie of modern industrial society; to provide a religious culture congruent to the needs of the material world of industrial capitalism; and, to provide a system of social security through Sunday schools, a network of credit, and mutual help that offered necessary, and essential, support to workers in the hostile environment of capitalist Britain.[18] These varied functions of evangelical nonconformity in general, and Methodism in particular, transformed what he called "religious deviance" into a "political safety valve" by allowing for the expression of "legitimate opposition to a reactionary regime" by way of the rejection of paternalism and deference.[19] At the same time, however, Methodism exercised "a form of social control over the political behavior of those attracted to it" by limiting the forms and character of political protest.[20] All in all, Gilbert posited Methodism as a force for "moderate radicalism" that "may have made an important contribution to public order and political stability."[21]

It would certainly be rash to discount the contributions of these historians to our understanding of the role and function of evangelicalism in the working-class community, but such broad pronouncements of "the impact

---

David Hempton, *Methodism and Politics in British Society, 1750–1850* (London: Hutchinson, 1984), 74–6.

[17] Hemptom, *Methodism and Politics*, 22–7, 77–80. Bernard Semmel argued similarly that Methodism was a theological counterpart to the liberal ideology of the so-called Age of Democratic Revolutions. However, Semmel, like Halévy, claimed that Wesleyan ideology prevented revolution by offering a "new, democratic faith in its Arminian Christianity, and by mobilizing popular energies in pursuit of personal salvation, while strengthening the motives for obedience and subordination." See, Semmel, *The Methodist Revolution* (New York: Basic, 1973), 192–3.

[18] A. D. Gilbert, *Religion and Society in Industrial England: Church, Chapel and Social Change 1740–1914* (London: Longman Group, 1976), 86–92.

[19] Alan D. Gilbert, "Methodism, Dissent and Political Stability in Early Industrial England," *Journal of Religious History*, 10, no. 4 (1979), 381–99.

[20] Ibid., 393.

[21] Ibid., 392.

of Methodism" are particularly inappropriate to the study of the northern miners for several reasons. In the first place, it must always be remembered that there was not one Methodism but several. Before 1840, while Wesleyan Methodism had a much broader institutional base than any other Noncon-formist sect, its impact, especially in times of social or economic crisis, was much less than its enthusiastic offshoot, Primitive Methodism.

Second, the attraction of individuals to Methodism was neither permanent nor was it necessarily long lasting in this era. Methodism's adherents con-tinually flowed into and out of the sect, in part reflecting the mobility of the population, in part reflecting the failure of Methodism to capture and retain its converts.

Finally, before 1840 it is certainly anachronistic to equate Methodism with the northern working class or even the northern mining population. Methodism was the religion of only a small portion of the working class, a portion, moreover, that was often viewed with suspicion by many workers. It was only the unique ability of some Methodists to transcend their sepa-rateness, their Weberian inner-worldly asceticism, while at the same time exploiting their peculiar qualities and achievements, that transformed certain members of the sect into working-class leaders. In sum, religion did not discipline, divert, or accommodate the northern working class to capitalism. That function was accorded to the workplace and the system of industrial relations in the northern coalfield.

The Established Church and the distribution of Nonconformity

By the first third of the nineteenth century, the Church in Durham epito-mized the weaknesses of the Church nationally. It was racked by pluralism and nonresidency, unresponsive to the changes in society brought about by the growth of local industry, and widely regarded as a haven for greedy and indolent clergymen. Moreover, support for the Church was not noticeably forthcoming from the local landed elites. Lord Londonderry, the arch-Tory magnate, was repeatedly angered and disaffected by the extraordinarily high renewal fines imposed upon the leases of his Church-owned mining prop-erties, while the Whig establishment, led by Earl Grey and J. G. Lambton, was vehemently against the Tory control of the Church in the North.[22]

The Church's position in Durham became so embattled by the 1820s that it went to the extent of suing a local newspaper editor for libel after a stinging attack upon the local clergy had been published. The editor of the *Durham Chronicle*, John Williams, in fact described the failures of the Church

---

[22] G. F. A. Best, *Temporal Pillars* (Cambridge University Press, 1964), 245–50.

with a fair degree of accuracy albeit with an unfair degree of hyperbole. It is the opprobrious conduct of the clergy, he wrote,

> which renders the very name of our Established Church odious till it stinks in the nostrils; that makes our Churches look like deserted sepulchres, rather than temples of the living God; that raises up conventicles in every corner, and increases the brood of wild fanatics and enthusiasts; that causes our beneficed dignitaries to be regarded as usurpers of their possessions; that deprives them of all pastoral influence and respect; that, in short, has left them no support or prop in the attachment or veneration of the people.[23]

The failures of the Church in Durham were rooted in both human avarice and institutional inertia. The Durham parish structure was simply too large and cumbersome to respond adequately to the increasingly rapid tempo of industrial development. William Cobbett, for one, was surprised to see so few churches in the North. He remarked that there were more church spires in any 20 miles in Wiltshire than in the 150 miles between Oldham and Newcastle.[24] The vicar of Pittington, a Durham mining parish, complained in a letter read before Parliament in 1836 that his parish was simply too large and populous to be adequately overseen. It contained, he estimated, six thousand souls spread over five thousand acres. Pittington needed a new chapel, the expansion of the existing church, and the establishment of at least two schools. "The case of Pittington," the vicar rightly concluded, "is the case, *mutatis mutandis*, of many other parishes in this diocese."[25]

Implicit in the remarks of both the editor of the *Durham Chronicle* in the 1820s and those of the vicar of Pittington more than a decade later is that the social significance of these large parishes lay in the clergy's manifest inability to minister to the social or spiritual needs of parishioners. In 1830, there were 71 parishes and extraparochial chapelries in County Durham. Their mean area was over 9,500 statute acres, and they had an average population of 3,669.[26] In comparison to many southern counties, these Durham parishes were certainly enormous. Norfolk, for example, was comprised of 731 parishes and Suffolk of 510.[27] In South Lindsey, studied by James Obelkevich, the average parish was a mere 1,736 acres, and only about a

---

[23] Quoted in ibid., 247. Williams's leader was provoked by the failure of the bishop and dean and chapter of Durham to mourn the death of Queen Caroline by ringing the bells of the cathedral. The libel case became a cause célèbre, and the editor was found guilty but went virtually unpunished.

[24] William Cobbett, *A Tour in Scotland and the Four Northern Counties* (London, 1833), 21.

[25] *Parliamentary Debates* (Hansard), vol. 33, third ser., House of Lords, 1183.

[26] John Bailey, *General View of the Agriculture of the County of Durham* (London, 1810), 3–7; *1831 Census: Comparative Account*, P.P. (1833), vol. 36, 85–91; *1841 Census: Enumeration Abstract*, pt. 1, P.P. (1843), vol. 22, 81–8.

[27] Ward, *Religion and Society in England*, 108.

third of all residents lived in parishes of more than 500 people.[28] J. R. Leifchild, the children's employment commissioner for north Durham and Northumberland, noted that in the mining regions, "some of the parishes … are so extensive as to defy adequate personal superintendence on the part of a single clergyman."[29]

The unfortunate legacy of Durham's ancient parish structure was compounded by failures of ecclesiastical policy and personnel. W. B. Maynard's work has revealed the extent to which pluralism and nonresidency affected the archdeaconry of Durham, that is, the area of the Durham diocese covering County Durham.[30] On average, about a third of Durham incumbents were nonresident in the first four decades of the nineteenth century, although this percentage declined from nearly 40 percent in 1814 to 20 percent in 1832. Furthermore, between 1774 and 1832, more than one-half of all Durham incumbents were pluralists. The greatest share of the responsibility for the persistence of pluralism in Durham, Maynard has argued, lay with the bishop of Durham and the dean and chapter of the cathedral. They not only tolerated pluralism but actively promoted it in part to augment poor livings, but more importantly to reward kin and connections.[31] Henry Phillpotts was perhaps the most notorious pluralist and nonresident of County Durham, but he was matched by the Reverend Edward South Thurlow, nephew of the lord chancellor, as well as by G. V. Wellesley, younger brother of the duke of Wellington.[32]

In marked contrast to the situation in the south of England, the patronage of Church livings in the northern coalfield was concentrated overwhelmingly in the hands of the bishop and the dean and chapter. Of ninety benefices listed in the 1810 agricultural survey of County Durham, fifty-seven, or over 63 percent, were in the patronage of the Church: thirty-one in the patronage of the bishop, seventeen in the patronage of the dean and chapter, and nine in the patronage of other ecclesiastics (three vicars, three rectors, two curates, and the archdeacon of Northumberland).[33] Only thirty-three livings (nearly 37 percent), were in the hands of nominally nonecclesiastical patrons: five benefices were in the patronage of the Crown, six were presented by university colleges and hospital corporations, and twenty-two were in the pa-

[28] James Obelkevich, *Religion and Rural Society: South Lindsey, 1825–1875* (Oxford University Press, 1976), 8–9.
[29] Leifchild, *Our Coal and Our Coal-Pits*, 216.
[30] I owe special thanks to W. B. Maynard who kindly lent me a draft of his article "Pluralism and Non-Residency in the Archdeaconry of Durham, 1774–1856: The Bishop and Chapter as Patrons."
[31] Ibid., 20.
[32] Ibid., 18, 27–8.
[33] Bailey, *General View*, 3–7.

tronage of the laity. This nearly inverts the findings of other local studies of Devon, Derbyshire, Oxford, and South Lindsey where ecclesiastical patrons in each case presented to less than 30 percent of the livings.[34]

The concentration of ecclesiastical power in the hands of the bishop and dean and chapter effectively excluded lay patrons; and, by this exclusion the Church alienated an essential source of political support from among the Durham gentry and aristocracy. At best, Durham landowners were ambivalent toward the Established Church. The necessity for some measure of reform was recognized by nearly everyone.[35] But at worst, landowners, particularly those involved in the coal trade, were overtly hostile to the Church and to the power it wielded through its possessions. The Church's extensive landholdings in north Durham made it the proprietor of several collieries and also gave it control of wayleaves to both the Tyne and Wear. Many coal owners leased collieries from the Church, but none were more prominent than Lord Londonderry, whose principal collieries were leasehold of the dean and chapter of Durham Cathedral.[36]

The renewal of the leases of Pittington and Rainton collieries continually brought Londonderry's political ideology into conflict with his economic self-interest. In 1829, he wrote bitterly and threateningly of the Church's policy of demanding large renewal fines upon the cessation of leases:

> If they [the dean and chapter] are *avaritious* the H[ouse] of Lords will ring with their avidity – ... the *Times* we live in and the March of Intellect will bear no *Rapacity* from the Church. Greediness might upset all the holds they have, and it is in their Interest to cultivate the Support of the Aristocracy of the Land ... And if they are unbending and overeaching, the Halls of Parliament, the Remote Parts of the Empire, shall hear of the Reverend Body in their true colours – ... depend upon it, the Grain is lighted in a manner that never before occurred, It only wants the Match, and I am not the Man to shrink from firing the Mine, if I have just provocation.[37]

Less than a decade later, in 1837, Henry Morton, Lord Durham's estate agent, reported to Earl Grey that it was common knowledge in the county that Londonderry had paid "upwards of £100,000 in the shape of fines to the Dean and Chapter of Durham."[38] Further, it was rumored that £90,000

---

[34] Maynard, "Pluralism and Non-residency," 10, records 67 percent of the livings in the county in ecclesiastical hands in 1774; see also Obelkevich, *Religion and Rural Society*, 112.

[35] See *Report of Speeches delivered at a Public Meeting of the Friends of the Established Church* (Newcastle-upon-Tyne, 1834).

[36] See Chapter 3.

[37] D.C.R.O., Londonderry Papers, D/Lo/B 303(24), Londonderry to Buddle, 11 July 1829.

[38] Grey Papers, Durham University, Department of Paleography and Diplomatic, Morton to Earl Grey, 20 May 1837.

of those fines had gone directly into the pockets of the prebends of the Church.[39]

The Church in the northern coalfield thus was racked by at least two fundamental problems. On the one hand, its administrative organization was inefficient, unable to respond to the changing social conditions, and corrupt. Reflecting upon the Church's efforts, one coal owner simply lamented: "The Church should have done her duty better towards" the miners.[40] On the other hand, the Church's immense wealth and economic power brought it into contention with the local gentry and aristocracy. At a County Durham reform meeting in 1831, Cuthbert Rippon, a local J.P., was applauded when he specifically linked Church reform with parliamentary reform. In both instances, he said, "the people, as yet, reconciled to the corruptions of the church and the state by the power of worn-out deception, now saw and knew the truth.... They detected the intent and purpose of our spiritual and secular establishments to be personal profit and worldly advantage."[41] Although the Church in the mining regions of the Northeast was not precisely moribund and it would later make substantial efforts to extend its influence among the miners,[42] spiritual life in the mining communities of the early nineteenth century was increasingly focused upon Nonconformist sects, particularly Methodism.

By 1830, Methodism was the dominant Nonconformist sect in Durham, a fact that is confirmed by a return of Nonconformist meetinghouses ordered by the Durham Quarter Sessions in 1829. In all, the surviving returns identified 213 chapels and meetinghouses. (Due to the loss of some returns, the official lists have been supplemented here by information collated from local histories. From these sources, 23 further chapels have been identified for the period before 1830.)[43] Table 6.1 reveals that Wesleyan Methodists registered nearly twice the number of chapels and meetinghouses than their closest rivals, the Primitive Methodists. Altogether, Methodist sects accounted for nearly seven of every ten Nonconformist chapels and meetinghouses in the county.

Numerically, Methodism's greatest strength lay in its ability to penetrate both town and country. This contrasts dramatically with Roman Catholicism,

[39] Ibid., 2 April 1837.
[40] *Reports from Commissioners: Children's Employment (Mines)*, P.P. (1842), vol. 16, 609.
[41] *Newcastle Chronicle*, 1 February 1831.
[42] See Brendan Duffy's contribution to "Debate: Coal, Class and Education in the North-East," *Past & Present*, no. 90 (February 1981).
[43] D.C.R.O., Return of Nonconformist Meeting Houses, 1829, Q/R/RM 2–15. The returns for Chester Middle Ward have not survived. They have been supplemented by William Fordyce, *History and Antiquities*, and Parson and White, *History, Directory and Gazetteer*.

Table 6.1. *Nonconformist chapels and meetinghouses in Durham, 1829*

Sect or denomination	Number of chapels or meetinghouses (%)	
Wesleyan Methodist	102	(43.2)
Primitive Methodist	56	(23.7)
Roman Catholic	20	(8.5)
Baptist and Particular Baptist	15	(6.4)
Independents	8	(3.4)
Presbyterian	6	(2.5)
Quaker	6	(2.5)
Methodist New Connexion	5	(2.1)
United Secession of Scotland	5	(2.1)
Established Church of Scotland	4	(1.7)
Unitarian	3	(1.3)
Congregationalist	2	(0.9)
Jewish	1	(0.4)
Other	3	(1.3)
Total	236 (100)	

Source: D.C.R.O., Return of Nonconformist Meetinghouses, 1829, Q/R/RM 2–15.

whose institutional strength was disproportionately concentrated in small towns and villages of under five hundred inhabitants, as well as with the so-called Old Dissent (Baptists, Presbyterians, Independents, Quakers, and Unitarians), whose meetinghouses and chapels were only rarely located outside of the county's largest towns.[44] Methodism, however, made a particularly dedicated effort to mission both the urban areas of early-nineteenth-century Durham and the small to medium-sized towns and industrial villages of the county.

Table 6.2 locates Wesleyan Methodist, Primitive Methodist, Roman Catholic, and Old Dissent meetinghouses and chapels with reference to the population distribution of the county circa 1831. This table should be read

---

[44] Rural Catholicism survived largely due to the support of a small number of gentry families in Durham among the most prominent of whom were the Claverings. See John Bossy, *The English Catholic Community, 1570–1850* (London: Darton, Longman & Todd, 1975), 301–2, 320–2. I have eschewed the invocation of the distinction between "open" and "closed" parishes advocated by Everitt and Obelkevich, among others, to explain the success or failure of nonconformity. In Durham, this distinction does not hold. Given the size and acreage of the average parish in the county, every parish was "open" and none, or very few, were "closed."

The struggle for market power

Table 6.2. *Distribution of Nonconformist meetinghouses, 1829*

	Size of towns, 1831 (% total pop.)				
Sect/denomination	<100 (1.3)	100–499 (13.7)	500–999 (8.8)	1,000–1,999 (13.7)	>2,000 (62.5)
Wesleyan Methodist (N = 102), %	—	29 (28)	19 (19)	12 (12)	42 (41)
Primitive Methodist (N = 56), %	—	11 (20)	12 (21)	5 (9)	28 (50)
Roman Catholic (N = 20), %	1 (5)	12 (60)	—	1 (5)	6 (30)
Old Dissent (N=38), %	—	3 (8)	1 (3)	5 (13)	29 (76)

Source: D.C.R.O., Return of Nonconformist Meeting Houses, 1829, Q/R/RM 2–15; *1831 Census:* Comparative Account of the Population of Great Britain in 1801, 1811, 1821, 1831, P.P. (1833), Vol. XXXVI, 85–91.

with some care since it takes no account of attendance or adherence but only of institutional presence in the community. Thus a chapel serving several hundred adherents in Sunderland, for example, appears equal to a meetinghouse serving a dozen or so persons in a rural village. Nonetheless, Methodism's ability to reach both urban and rural residents of the county is quite noticeable. Indeed, this was one of Methodism's most notable achievements in the North and what distinguished it in part from earlier Nonconformist movements.

One should be wary, therefore, of broad generalizations regarding the geographic distribution of Nonconformity, particularly Methodism. Hobsbawm's contention that Primitive Methodism was "pre-eminently a *village* labour sect" (his emphasis) because industrial towns and large cities were "inhospitable to working-class religion" is certainly inapplicable.[45] By 1830, one-half of all Primitive Methodist meetinghouses and chapels were located in Durham's largest towns and cities where over 60 percent of the people of the county lived. Comparatively, Wesleyan Methodism was statistically overrepresented in rural Durham. Yet over 40 percent of its chapels and meetinghouses were also located in the most highly urbanized areas of County Durham. This pattern of the geographical distribution of Methodism is particularly important for the study of the attraction and influence of evangelical ideology on working-class movements; for, as will be argued, there was little or no correlation (whether direct or inverse) between the oscillations of general Methodist membership and broad-based political activism. Meth-

[45] E. J. Hobsbawm, *Primitive Rebels* (New York: Norton, 1959), 137.

odist culture helped to form an elite cadre of trade union leaders, but mass politics and organized religion were only tenuously connected.

Methodism and its adherents

Methodism first took root in the northern mining districts during John Wesley's mission to the region in the early 1740s. By the 1780s, there were just over a thousand Methodists in the Sunderland Circuit of north Durham, which covered the mining regions roughly between the Tyne and Wear as well as Durham City and Sunderland.[46] The fragmentation of Methodism that followed Wesley's death in 1791, however, eventually led to the founding of a number of related sects, including the Primitive Methodists in 1810.[47] The first Primitive Methodist missionaries crossed into Durham in 1820, and as we have seen, before the end of the decade they became the second largest Methodist sect in that county. Much of what follows will focus on the sect of Primitive Methodism. While this emphasis may seem misplaced with regard to the numerical relationship of Wesleyans and Primitives, the latter were by far the most active and conspicuous of all Nonconformist groups during this period. Their particular significance in the history of northern labor relations lay in the fact that Primitive Methodist lay preachers led the regional union movements of the 1830s and 1840s. Later in the century, Primitive Methodists would continue to exert a profound influence upon northern politics and industrial relations: John Wilson would help found the Durham Miners' Association, and Thomas Burt of Northumberland would be one of the first working-class members of Parliament. An analysis of the role of religious ideology in the development of industrial relations must therefore pay particular attention to Primitive Methodism.

The split between the Primitive Methodists and mainstream Methodism was not based upon theological principle, but was largely a matter of style, discipline, and politics.[48] Hugh Bourne and William Clowes of Staffordshire emerged as leaders of a group of local Methodists encouraged by the American revivalist Lorenzo "Crazy" Dow, who first preached at Liverpool in 1805. Dow's open-air camp meetings and uninhibited revivalist manner drew the attention not only of disenchanted Methodists but of national Methodist officials as well as Anglican bishops. As early as 1807, the national Methodist Conference had denounced open-air camp meetings as "highly improper in England and likely to be productive of considerable mischief."[49]

---

[46] Colls, *Pitmen of the Northern Coalfield*, 146–8.
[47] Hempton, *Methodism and Politics*, 55–84.
[48] Ibid., 92–104.
[49] Quoted in H. B. Kendall, *The Origin and History of the Primitive Methodist Church*, 2 vols.

Furthermore, Lord Sidmouth's attempt between 1809 and 1811 to control itinerant preaching, one of Wesley's cherished legacies, forced the national Methodist leadership to portray itself, in Hempton's words, as "Anglican in sympathy, Protestant in character, disciplined in ecclesiastical organization, and sustainer of a stable social order" in order to protect itself from state control.[50] Eventually, Methodist pressure along these lines helped to secure from Lord Liverpool's government the passage of the 1812 Toleration Act as well as the repeal of the Five Mile Act and the Conventicle Act.[51] Within Methodism, therefore, the Primitives came under attack precisely because they threatened the internal discipline and political facade of Methodism. As Wesleyan Methodism became increasingly orthodox and establishmentarian after 1820, it also increasingly relinquished its ability to influence the culture and politics of the English working class,[52] a role that in the northern coalfield eventually fell to the Primitive Methodists.

As was the case with Hugh Bourne and William Clowes, Primitive Methodism's initial attraction was the enthusiasm unleashed at open-air camp meetings and love feasts. The sect became known commonly as the "Ranters" because their evangelical style encouraged the physical expression of the holy possession of the spirit through speaking in tongues, uncontrolled sobbing, wild dancing, and fainting.[53] Early Primitive Methodist sects were probably made up largely of disenchanted Wesleyans. Joseph Peart, a schoolmaster at Chirton and former Wesleyan preacher, invited William Clowes to North Shields in 1821 because he was dissatisfied with the moderation shown by local Wesleyans.[54] Similarly, Joseph Spoor and his sister Jane, famous local preachers in the Newcastle area, were originally Wesleyans "but had found the spirit of the early Primitives more in keeping with their fervent souls."[55]

Thus, the early links between Wesleyan and Primitive Methodism were necessarily quite strong at the local level. Something akin to a joint revival took place at Haswell Colliery in 1838, where the famous Wesleyan preacher

(London: Dalton, 1909), vol. i, 77; see also John Petty, *The History of the Primitive Methodist Connexion* (London: 1860), 23.

[50] Hempton, *Methodism and Politics*, 102.

[51] Ibid., 98–104; Ward, *Religion and Society in England*, 59–62; 85–9.

[52] Hempton, *Methodism and Politics*, 110; see also David Thompson, *Nonconformity in the Nineteenth Century* (London: Routledge & Kegan Paul, 1972), 64–5, and Robert Currie, *Methodism Divided: A Study in the Sociology of Ecumenicalism* (London: Faber, 1968).

[53] John Kent, *Holding the Fort: Studies in Victorian Revivalism* (London: Epworth, 1978), 38–70.

[54] W. M. Patterson, *Northern Primitive Methodism* (London: Dalton, 1909), 334; Kendall, *Origin and History*, vol. 2, 169.

[55] Patterson, *Northern Primitive Methodism*, 188.

Peter Mackenzie was saved.[56] At Ramshaw on Tyneside, the Primitives and Wesleyans shared a meetinghouse owned by the local lead mining company.[57] In fact, up to 1830 it is clear that Primitive Methodism made very little headway outside of areas already missioned by the Wesleyans. In the 1829 survey of Nonconformist meetinghouses, there were only four reported locations in County Durham where a Ranter meetinghouse or chapel had been established without a corresponding Wesleyan institution. Conversely, the Wesleyans had established a substantially wider network of chapels and meetinghouses virtually without competition from other sects. There were at least twenty-three Wesleyan congregations reported in towns and villages without any other Dissenting institutions.[58] Proportionately, while 22 percent of Wesleyan establishments reported in 1829 were located in towns or villages without a corresponding Primitive Methodist group, only 7 percent of the Primitive Methodist meetinghouses and chapels were situated similarly.

The close connection between Wesleyan Methodism and the early Ranters was also apparent in the occupational makeup of the founding members of the new sect in the Northeast. The earliest Primitive Methodists frequently reflected Wesleyan respectability. John Robinson, one of the first class leaders in South Shields in 1822–3, was a shipowner living on East King Street. He was able to lend the connexion £460 to build the first Primitive Methodist chapel in that town.[59] Thomas Davison, an early chapel steward in South Shields, was a grocer, and two other early members, William Hardy and Joshua Shaw, were a publican and grocer, respectively.[60] In Sunderland, there appear to have been similar ties to commercial, mercantile, and artisanal interests among the early Ranters. Henry Hesman was an insurance broker at Thornhill's Wharf; John Cooper was a fitter in Upper Sans Street; and William Brass was a boot- and shoemaker working on the High Street.[61]

Outside of these large towns, the occupational makeup of early Primitive Methodism was likely to be more mixed but still not exclusively working class. In Shotley Bridge near the Consett Ironworks, for example, Robert Taylor, a yeoman, and William Robson, a grocer, conveyed land to the Primitive Methodists on which was built the sect's second chapel in County Durham in 1822. Of fourteen other members of that early chapel for which

[56] Ibid., 287.
[57] Ibid., 191.
[58] D.C.R.O., Return of Nonconformist Meeting Houses, 1829, Q/R/RM 2–15.
[59] Patterson, *Northern Primitive Methodism*, 229–30; Parson and White, *History, Directory and Gazetteer*, vol. 1, 290.
[60] Patterson, *Northern Primitive Methodism*, 231–3; Parson and White, *History, Directory and Gazetteer*, vol. 2, 666.
[61] Patterson, *Northern Primitive Methodism*, 249; Parson and White, *History, Directory and Gazetteer*, vol. 1, 355–6, 375; vol. 2., 667.

occupational information survives, there were three shoemakers, two joiners, two tailors, a yeoman, a mason, a clogger, a millwright, a sword grinder, and a miner. They were, moreover, led by John Dover Muschamp, a local gentleman.[62]

The variety of connections between Wesleyanism and early Primitive Methodism reinforces the conclusion that the split between the two were originally matters of style, discipline, and politics, not theology. It also serves to point out part of the reason that despite the fact that Primitive Methodism later dominated many working-class communities in the northern coalfield, Ranterism retained its emphasis on orderliness and respectability. The division between Wesleyanism and Ranterism was fundamentally a question of the techniques of salvation, not the path of earthly conduct. For the Primitives, this technique included spiritual ecstasy and open-air revivalism.

Once saved, therefore, the conduct of a Primitive differed little from a Wesleyan. Popular culture was replaced by the trappings of respectability. The pitman's posey jacket, purple stockings, shoulder-length hair, and beribboned hats gave way to the black suit, top hat, and white shirt worn by the Primitive Methodist lay preacher and miners' leader Tommy Hepburn during the pitmen's strike of 1831–2. Primitive Methodist open-air camp meetings were organized specifically to counter the evil influences of popular "wakes." Hugh Bourne, in fact, had hoped originally to hold the famous Mow Cop meeting in late August in order to "engage our young members, and preserve them from being seduced by the vanities of the wake."[63] But other Methodists sought to avoid a conflict with the annual wake and decided to hold their meeting as early as possible. Thus the first Primitive Methodist meeting was held on 31 May 1807.[64]

These attacks upon popular culture brought the sect notoriety and heralded a degree of cultural transformation. At South Side, in Durham, "centres of gambling were broken up; confirmed gamblers burnt their dice, cards, and 'books of enchantment'; drunkards, hopeless, incurable sots, were freed from the dread tyranny of fiery appetite; pugilists, practised and professional, and cock-fighters of terrible experience, turned from their brutalities."[65] In Newcastle, where the miners and keelmen were "proverbial for drunkeness and its attendant vices," according to W. Lister, a local Ranter preacher, a revival in 1822 resulted in "a general reformation of manners. Sobriety, industry, and peaceable behaviour took the place of drunkeness,

---

[62] Patterson, *Northern Primitive Methodism*, 210.
[63] Quoted in Petty, *History*, 14.
[64] Petty, *History*, 14; Kendall, *Origin and History*, vol. 1, 62.
[65] Patterson, *Northern Primitive Methodism*, 70.

Table 6.3. *Wesleyan converts, 1825–9 (* N = *184)*

	Male	Female
Sex (*N* = 183), %	37.2	62.8
Marital status (*N* = 124), %		
Single	43.2	70.0
Married or widowed	56.8	30.0
Average age (*N* = 178), years	30.0	26.2
Age × marital status (*N* = 149), years		
Mean age of singles	20.8	19.9
Mean age of those married or widowed	37.7	37.8

Source: T.W.C.A., 1041/3, Brunswick Methodist Chapel, Register and Minute Book, 1825–35.

indolence, and brawls and contentions."[66] Another lay preacher, Thomas Batty, who worked among the lead miners of Weardale, congratulated himself for making enemies of publicans and friends of tailors and women:

> We had many enemies of one kind at first, but now their mouths are stopped, and we have got enemies of another kind – publicans, because their custom is lessened, many drunkards having become sober. On the other hand, we are getting a few friends among the tailors, some persons who formerly went in rags being now able to get new clothes, and we have many friends among the women, whose husbands were drunkards, spending much of their time and money in public houses, but who, having become sober, they have now the comfort of their company at home, and the pleasure of going with them to the house of God.[67]

While the friendship of tailors may be difficult to substantiate, an analysis of Wesleyan Methodist converts in Newcastle between 1825 and 1829 confirms the important, indeed preponderant, role of women. Although the numbers are too small to establish a high degree of certainty, of 183 converts, nearly 63 percent were women[68] (Table 6.3). Moreover, reflecting information calculated by Malmgreen for the Macclesfield area, a high proportion of these women were single.[69] In this case again, numbers are very small: The

[66] Petty, *History*, 145–6.
[67] Ibid., 147.
[68] For this and the following data, see T.W.C.A., 1041/3, Newcastle and Brunswick Central Circuit, Brunswick Methodist Chapel, Register and Minute Book, 1825–1835.
[69] Cited in Hempton, *Methodism and Politics*, 13.

marital status of only 80 of 115 women has been recorded. Nonetheless, 56 of those women, or 70 percent, were unmarried. Comparatively, the marital status of 44 men was recorded of whom only 43 percent were unmarried. There is, therefore, reason to believe that Methodism particularly attracted women, and single women to a greater degree than married women.

It may very well have been that Methodism provided both community and companionship for women who in the Northeast, as elsewhere, were increasingly segregated from both the workplace and the leisure activities of men. Men, however, were more likely to be married upon entrance into the Methodist sect. And they may have considered Methodism either as an aspect of new family formation that connoted seriousness and responsibility or as a group to turn to after they had passed middle age, when their earning potential began to decline and when they were faced with an insecure future.

The surviving data from Wesleyan converts make such hypotheses tenable because they also record the age of the new converts as well as their marital status. When the two are correlated, it is clear that Methodism's appeal acted most profoundly upon both men and women at particular passages in their lives. Generally, the average age of new members was higher for men than for women: Men entered the Wesleyan sect on average at age thirty while women were recruited at an average age of twenty-six.

Methodism exerted its greatest attraction at two points in people's lives: either when they first entered adulthood around age twenty-one or when they began nearing age forty. Thus, the mean age of single women (who made up 70 percent of all female converts) was just under twenty, while that of single men was just under twenty-one.[70] Methodism seems to have lost its attraction for both men and women after their late teens and early twenties, however, perhaps because work and family tended to marginalize worship. Yet members of both sexes apparently sought out the Methodists again after their midthirties, that is, after families had been established and working-class parents became less sanguine about their future. The mean age of married or widowed women and men who entered the sect was just under thirty-eight years old.

While converts to Methodism tended to be drawn from two groups, young singles under age twenty-one and older married adults over age thirty-five, there was, nonetheless, a marked difference between men and women with regard to the timing of their entrance into the Methodist sect. Women had a far greater tendency than men to join the Methodists when they were

[70] This accords with Deborah Valenze's findings for southwest England between 1827 and 1841 where the median age of female preachers was nineteen and a half: Deborah Valenze, *Prophetic Sons and Daughters: Female Preaching and Popular Religion in Industrial England* (Princeton: Princeton University Press, 1985), 112.

young and single, while men had a greater tendency to join when they were older and married. Relatively few married or widowed women converted to Methodism, 30 percent of all female converts, while a majority of men, nearly 57 percent, were either married or widowed.

While this data provides interesting insights into how Methodism may have been perceived by men and women, it seems to be generally true that initial conversion, based as it was on immediate religious experience, was less a problem to the sect than was the long-term retention of members. Retention required, after all, a "general reformation of manners," and the conflict that ensued within each individual between the willingness of the spirit and that of the body was not one easily resolved. Jabez Bunting had noted the phenomenon in the case of West Riding Methodism: Progress there had been "more swift than solid; more extensive than deep."[71] Moreover, in the northeastern coalfield, the personal crisis attendant upon Methodist discipline was compounded by the migration patterns of miners and their families. Methodist classes were in a constant state of flux as members entered and left both the sect and the community. Further, natural disasters, especially the cholera epidemic of 1832, tended briefly to attract members who later fell away.[72]

Methodist membership was by nature a transitory and temporary experience for a significant number of adherents. Contrary to Gilbert's contention that movement into and out of Methodist sects was unlikely and that "the overall incidence of defection was low" because Methodism was one of several "totalitarian religious cultures," evidence from the Northeast points to a rather constant stream of both new members and fallen brethren.[73] At the level of the individual chapel community, Primitive Methodist membership in the collieries could be highly volatile. For example, at Jarrow there were thirty-one full members of the Primitive Methodists and four trial members in March 1833. At the next quarter, June 1833, six full members had moved to another circuit and two other full members had left the sect. By the following September, two further full members had also left. At Hebburn in March 1832, in one of three Primitive Methodist classes there, ten full and thirty-six trial members had been reduced within three months by the removal of one member to another circuit and the falling away of fourteen trial and three full members.[74] At Lumley, the site of one of Lord Durham's collieries, membership totaled ten Ranters in June 1829,

[71] Quoted in Hempton, *Methodism and Politics*, 104.
[72] R. B. Walker, "The Growth of Wesleyan Methodism in Victorian England and Wales," *Journal of Ecclesiastical History*, 24, no. 3 (1973), 270–1.
[73] Gilbert, "Methodism, Dissent and Political Stability," 397–8.
[74] T.W.C.A., 1096/23, South Shields Circuit: Quarterly List of Members, 1832–1841.

Table 6.4. *Total membership in South Shields Primitive Methodist*
Circuit, 1832–40

Year	Total membership	Percentage of change
1832	800	
1833	472	− 41
1834	410	− 13
1835	490	+ 19.5
1836	442[a]	− 9.8
1837	420	− 4.9
1838	468	+ 11.4
1839	533	+ 13.8
1840	619	+ 16.1

[a]Estimated figure.
Source: T.W.C.A., 1096/23, South Shields Circuit: Quarterly List of Members, 1832–41.

rose to eighteen members by December of the same year, and then fell back to ten members by December 1830.[75]

A similar pattern of additions and defections from the sect can also be discerned at the circuit level. In the South Shields Primitive Methodist Circuit, total membership fell from a peak of 800 in 1832 to 420 in 1837 and then began to rise again to 619 by 1840 (Table 6.4). In this particular case, the statistics of total membership have the even greater advantage of including what the Primitive Methodists called "removals," or, the relocation of followers from one circuit to another, as well as those who had left the sect or had "fallen away," that is, ceased to attend class meetings. By correcting for removals, the effect of migration on one circuit's membership can be distinguished to some extent from a real decline in membership. When this is done, an average of 12 percent of total members annually left or fell away from the sect in the South Shields Circuit between 1832 and 1838 (Table 6.5). At some points, the defection rate from the South Shields Circuit reached nearly a quarter of all members, as it did in 1836–7. Yet as with other membership data, these numbers must still be used as only a rough guide to trends rather than as an exact accounting. For example, despite the fact that there was a 41 percent drop in South Shields membership in 1832–34, circuit records account for only about two-thirds, 210 of 328, of the full and trial members who left, fell away, or moved from the circuit. Thus the actual defection rate was likely to have been significantly

[75] D.C.R.O., M/Du 34, Primitive Methodist Circuit Accounts, 1828–1836.

Table 6.5. *Members "left" and "fallen": South Shields Primitive Methodist Circuit, 1832–38*

Year	Total membership	Members left and fallen	Percentage
1832–3	800	85	10.6
1833–4	472	21	4.5
1834–5	410	25	6.1
1835–6	n.d.	—	—
1836–7	420	104	24.8
1837–8	468	64	13.7

Source: T.W.C.A., 1096/23, South Shields Circuit: Quarterly List of Members, 1832–41.

higher than that indicated only by the members classified as left or fallen. The extent to which these people should be considered among the less conscientious "penumbra of adherents," as Gilbert called them, rather than full and dedicated members, may be open to argument. Yet there is little reason to dispute the assertion that the Methodist community was an active and volatile one and that class leaders, like Tommy Hepburn, who left the Primitive Methodists at the end of the 1831–2 strike, as well as regular members easily fell away from the movement.

Methodism and trade unionism

The mass trade union movements of the 1820s through the 1840s in the northern coalfield were not motivated by religious ideology. This may seem to be a difficult proposition to defend given the fact that the most well known miners' unions of the first half of the nineteenth century, Tommy Hepburn's union of 1831–2 and the National Miners' Association of 1842–4, were led by a preponderance of Primitive Methodist lay preachers. Yet Primitive Methodist trade union leadership rarely was translated into an evangelical revival. While it is true that Methodist leadership was a function of their skills as bargainers, orators, organizers, and as literate members of the community, it certainly is not true that the success of Methodist leaders was based upon their ability to conflate religious with political mobilization, as many have argued.[76] In fact, Methodist trade union leadership was accepted by the rank and file largely because the former were able to transcend their

[76] For just one example, see Robert Currie, Alan Gilbert, and Lee Horsley, *Churches and Churchgoers: Patterns of Church Growth in the British Isles since 1700* (Oxford: Clarendon, 1977), 109.

separateness, their inner-worldly asceticism, and put their skills to work for the broader community.

The Primitive Methodist community certainly offered an alternative to the traditional "rough" culture of the northern miners for the "serious" individual. John Wilson, the famous Durham trade unionist of the second half of the nineteenth century, was struck by how his conversion to Methodism alienated him from his old mates. They could not believe he would pass up a friendly game of cards, and he stopped drinking and gambling with them as well.[77] Miners tended to be suspicious of the Methodist attraction to books and desire for education. Jack (later Lord) Lawson remembered that his reading separated him from "the rough and ready life" and that to the people around him he "was simply voted 'queer.' "[78] (Coincidentally, Hugh McLeod has shown that in working-class London of the late nineteenth century, those who took an interest in books and learning were considered "pushy" and "putting on airs" by the rest of the community.)[79]

To many with such interests, local culture and society provided few alternatives, and they were thus drawn to the company of the Methodists. "Looking back now," Lawson wrote, "I see that it was inevitable that I should ultimately seek the company of the serious-minded people who gravitated together and formed the [Methodist] 'Society.' " Lawson concluded: "I was given their warm, helpful friendship and the hospitality of their homes. No longer was I 'queer' or 'alone.' "[80]

This community of like-minded people formed the core of both early trade union leadership in the Northeast and Primitive Methodism. Significantly, in many biographies of nineteenth-century miners' leaders, religion and theology are decidedly less important than the sense of community and discipline. Moreover, to many miners' leaders, just as to many "ordinary" Primitive Methodists, the attraction of the sect was often both fleeting and tenuous. Lawson wrote copiously of the camaraderie of the Methodist Society but hardly a word on theology or his conversion. Thomas Burt, a Northumberland miner and one of the first working-class M.P.s, left the Primitive Methodists and adopted Unitarianism. And, perhaps most significantly, Tommy Hepburn, the leader of the Great Strike of 1831–2, left the Primitive Methodists after the strike's failure.[81]

---

[77] Wilson, *Memories*, 202–3.

[78] Lawson, *A Man's Life*, 107–8.

[79] Hugh McLeod, *Class and Religion in the Late Victorian City* (Croom Helm, 1974), 48–55; see also Watson, *A Great Labour Leader*, 67, and Wilson, *Memories*, 206–8, for similar experiences.

[80] Lawson, *A Man's Life*, 107–8, 114.

[81] Lawson, *A Man's Life*; Watson, *A Great Labour Leader*, 165–71; John Oxberry, *Thomas*

Clearly, the attraction of Primitive Methodism went far beyond the theology of conversion and personal regeneration that was characteristic of this enthusiastic sect. The attraction of the Methodist community included an educated, literate, and respectable vision of working-class culture as opposed to the "rough and ready life" of the majority of miners and their families.[82] It is not surprising, therefore, that in times of social and economic distress, other members of the mining community tended to look toward the Methodists when their skills or attributes were needed. This was certainly true for later miners' leaders. John Wilson, the founder of the Durham Miners' Association, recalled that he had no desire to participate in union activities, but in the 1860s, he was pushed by his mates to become the local secretary because it was well known that he could read and write.[83] George Johnson, the viewer at Willington, Heaton, and Burdon Main collieries, identified the Methodists as the "educated people." In 1841, he explained to a parliamentary commission:

> In ordinary matters education makes no difference to them as workmen; but in strikes, etc., these educated people, or Methodists, are most decidedly the hardest to deal with. This is not always from taking the correct view of the nature of the disputes, but from self-sufficiency and self-satisfaction with their superiority.[84]

Robert Lowery, the radical and later Chartist who worked with Tommy Hepburn and the 1831–2 miners' union, similarly believed that leadership in trade union matters was often forced upon "men of good character":

> The mass of workers instinctively think the men of good character will manage their affairs best, and thus all of them unite to force the leadership upon these men. They are urged to it as a duty they owe their class, and accused of

---

Hepburn of Felling: What He Did for the Miners (Felling-on-Tyne: R. Heslop, 1959); *Primitive Methodist Magazine, 1865*, 546–7; Colls, *Pitmen of the Northern Coalfield*, 352.

[82] Interesting parallels may be drawn to the marked divergence between the "clerical prescription" of the social impact of conversion and its "lay perception," as described by Ned Landsman. In the case of the Cambuslang Revival of 1742–5, Landsman argues, the converts' sense of sin was not linked to the specter of Hell or the fear of damnation, but to the failure to live up to a set of values that included "literacy, sobriety, mutuality and assertiveness." See Ned Landsman, "Evangelists and their Hearers: Popular Interpretation of Revivalist Preaching in Eighteenth-Century Scotland," *Journal of British Studies*, 28, no. 2 (1989), 120–49.

[83] Wilson, *Memories*, 146. Wilson refused to become branch secretary of his union at that time stating, "I have always held the view that in the selection of lodge officials something more than mere clerking ability or power to declaim are essential." In a similar experience, Wilson was put forward by his mates to be the librarian of Haswell Colliery's workingmen's institute in 1869 because "there was no one so well qualified as John Wilson." See *Memories*, 224.

[84] *Reports from Commissioners: Children's Employment (Mines)*, P.P. (1842), vol. 16, 568.

cowardice or of leaning to treachery and to the side of the masters if they do not.[85]

This special relationship between the Primitive Methodist leadership and the rank-and-file union movement based upon the presumption of good character and education, not religious ideology, explains the odd dysfunction between religion and working-class leadership in the first half of the nineteenth century. That is, while there was very little correlation between the incidence of revivalism and trade unionism in the northern coalfield, the defeat of union movements almost inevitably contributed to a decline in Methodist membership. The unique connection can clearly be seen in the case of Tommy Hepburn's miners' union of 1831–2.

Tommy Hepburn was a Ranter preacher and hewer at Hetton Colliery, the center of the union movement. At least two other leaders of the union, John Johnson and Charles Parkinson, were also active in local Primitive Methodist societies.[86] The first mass rally of miners occurred on 26 February 1831, and the strike began at the conclusion of the previous bond on the fifth of April.[87] The coal owners, led by Lord Durham and Lord Londonderry, broke ranks in mid-May, and by the beginning of June, the miners' union had scored a striking victory. The miners' success lasted throughout the period of the succeeding bond, but in April 1832 they found themselves locked out by an angry and resolute ownership. By the late summer of 1832, the union was in disarray and near collapse. In September, the miners at Hetton Colliery sued the owners for work; this effectively signaled the defeat of the movement.

The Primitive Methodist leadership of the strike initiated some small-scale, localized religious revival. Again, although membership records cannot provide an absolute guide to Methodist influence and activity, the meticulousness and care with which these records were compiled can certainly serve as a rough measure of the fluctuations in the fortunes of Methodist classes and circuits. Not surprisingly, membership in the Hetton Primitive Methodist Society rose substantially.[88] In December 1830, the Hetton con-

---

[85] Harrison and Hollis, *Robert Lowery*, 75. Lowery, however, was wrong to ascribe union leadership in 1831–2 to the Wesleyans. Interestingly, in a nearly identical manner, Gravener Henson, in his own words, was "compelled" to assume the leadership of a nascent union among the Nottingham lace workers in 1809. Church and Chapman surmise that his reputation for literacy, education, and studiousness were among the qualities that commended him to other workers. See Roy A. Church and S. D. Chapman, "Gravener Henson and the Making of the English Working Class," in E. L. Jones and G. E. Mingay (eds.), *Land, Labour and Population in the Industrial Revolution* (London: Arnold, 1967), 133–4.

[86] Colls, *Pitmen of the Northern Coalfield*, 189–92.

[87] *Newcastle Chronicle*, 5 March 1831; see Chapter 7.

[88] D.C.R.O., M/Du 34, Primitive Methodist Circuit Accounts, Durham 1828–1836.

nexion reported fifty-one members and no persons on trial for admission into the sect. In March 1831, there were fifty-two members and thirty-four persons on trial. This was one month before the beginning of the strike. By June 1831, the first fortnight after Lord Londonderry's defection, Primitive Methodist membership at Hetton had jumped to seventy-five members with six more persons on trial. A similar growth in membership appears to have taken place in the vicinity of Lord Londonderry's own collieries. The Pittington connection of Primitive Methodists doubled from fourteen members in December 1830 to twenty-eight members in June 1831.[89]

However, outside of these areas the sect made little or no progress in saving souls. If the success of Primitive Methodist leadership was actually linked to their ability to translate political protest into religious enthusiasm, then one would expect membership to have risen precisely at the point of union victory.[90] However, this was not the case. After declining by nearly 3 percent in the first quarter of 1831, total membership in the Sunderland Primitive Methodist Circuit, which included the major centers of union activity such as Hetton, Lumley, Rainton, and Pittington as well as the major urban centers of Durham such as Sunderland, Monkwearmouth, Durham City, and Hartlepool, rose again by a bit less than 3.5 percent in the second quarter of 1831 when the strike began. However, the third quarter of 1831, the time of union victory, did not bring similar rewards to the Primitive Methodists. Between June and September 1831, membership declined again by another 3.5 percent.[91]

Moreover, it does not appear that union activism was linked specifically to Primitive Methodist camp meetings or large-scale revivals during the strike. Although counterfactual evidence is the most difficult to marshal, there are no records of large-scale camp meetings being conducted during the first summer of the miners' strike. A local preacher, J. Petty, found the mood of County Durham "sullen" during the summer of the strike in 1831, and his efforts to evangelize the collieries were a failure.[92]

However, while camp meetings and love feasts may have been the most conspicuous evidence of Primitive Methodist revivalist activity, they were not the only means of attracting converts to the sect. It is possible that the Primitive Methodist trade union leadership might have translated union meetings into revivalistic congregations. And there is some slight evidence that local trade union meetings were tinged with the evangelical spirit. Henry

---

[89] Ibid.
[90] Currie, Gilbert, and Horsley, *Churches and Churchgoers*, 108–9.
[91] D.C.R.O., M Du 34, Primitive Methodist Circuit Accounts, Durham, 1828–1836; see also Colls, *Pitmen of the Northern Coalfield*, 147–61.
[92] Quoted in Colls, *Pitmen of the Northern Coalfield*, 151.

Morton, Lord Durham's agent, was convinced that "the worst feature in this business that I see is that those Ranter Preachers are endeavouring to excite religious enthusiasm among the men."[93] But John Buddle, Lord Londonderry's viewer, never blamed the Primitive Methodists for contributing to trade unionism. Early in the course of the strike, Morton noted that the pitmen at Lumley Colliery began their union meetings with a prayer "to strengthen and prosper the cause" and that they had struck a banner emblazoned with "Stand fast therefore in the liberty wherewith Christ has made us free; and be not entangled again with the Yoke of Bondage: Galatians 5 chap, 1 verse."[94] There seems little doubt that religion was serving the cause of the strike here rather than vice versa. While references to "freedom" and "liberty" might connote spiritual as well as political values, the exhortation to stand fast must have had a particular resonance to those on strike.

The most striking evidence that Primitive Methodist trade union leadership was not the principal vehicle for an evangelical revival in the Northeast are the speeches of the union leaders that were reported in the local press. Tommy Hepburn never resorted to a frenzied exposition of the Gospel to generate support or enthusiasm. Hepburn repeatedly emphasized the importance of "good conduct," respect for the law,[95] "peaceable demeanour,"[96] mutuality, and perseverance.[97] In March 1832, in fact, Hepburn "forcibly pointed out the advantages of truth and reason [to a mass rally of the pitmen], and urged them to an adherence of these."[98] In the same vein, Benjamin Pyle, another union leader, offered a resolution to be adopted by a general meeting of the pitmen that "we henceforward persevere in the ways of truth and reason."[99]

However, this is not to deny that religious rhetoric and symbolism were entirely absent from all general union meetings. John Johnson and Charles Parkinson relied heavily on biblical references in their public speeches. Johnson, for example, once compared the pitmen's circumstances to "the Israelites under the hard-hearted Pharoahs of old." Just as the Red Sea was a barrier to the freedom of the Israelites, he said, so "there might appear a barrier to stand before them – the barrier of poverty."[100] Similarly, Parkinson,

---

[93] Lambton Mss., Morton to Durham, 16 April 1831. See also Morton's letter to the Home Office, P.R.O., H.O. 40/29/89, 8 June 1831.
[94] Lambton Mss., Morton to Durham, 16 April 1831.
[95] *Newcastle Chronicle*, 2 June 1832.
[96] Ibid., 20 August 1831.
[97] Ibid., 23 June 1832.
[98] Ibid., 10 March 1832.
[99] Ibid., 10 March 1832.
[100] Ibid., 21 April 1832.

who was once referred to as "that slick headed ranting knave"[101] in one of Buddle's few derogatory references to Primitive Methodism, hoped to establish a "general sinking fund for all classes" so that all might "live in union as men and Christians, dispensing and encouraging moral knowledge, and tending to evangelise the world."[102] However, it would be extremely difficult to discern the rhetoric of Primitive Methodist fundamentalism from that of a discourse based on a generally accepted Christian culture and heritage without further evidence of a decided convergence of revivalism and trade unionism.[103]

In 1831, there was, therefore, only a small degree of coherence between the success of temporal radicalism and evangelicalism. A small Methodist revival did coincide with union activism, but it was extremely localized and contained. Generally, Primitive Methodist leadership of the trade union movement in Durham and Northumberland did not translate directly into religious enthusiasm or wide-scale revivals. In the short term, neither strike activity, union organization, nor union victory had a pronounced effect on chapel membership.

Methodist membership, however, did not maintain that rather simple correlation throughout the remainder of the strike. By the end of 1832 and the defeat of the union, membership in the Sunderland Circuit had risen by over 33 percent compared with the previous year. The coalfield was gripped by a pronounced religious revival. The otherwise reticent circuit accounts recorder was moved to comment on the dramatic growth of members on trial for admission to the sect in the first quarter of 1832: "Increase 625 this quarter," was the simple, but unique note.[104]

The precise timing of the revival indicates that the appearance of this revival was not a chiliastic response to the defeat of the union movement. Instead, as Walker has noted for Wesleyan membership, the onset of the cholera epidemic of 1831–2 sparked a religious revival.[105] The cholera first appeared in the Northeast during the final weeks of October 1831. Sunderland was the scene of the first confirmed cholera death in England in

---

[101] D.C.R.O., Londonderry Papers, D/Lo/C 142(851), Buddle to Londonderry, 21 June 1832.
[102] *Newcastle Courant,* 21 April 1832.
[103] A similar point is made by James Epstein in regard to Primitive Methodist leadership in the Chartist movement at Nottingham. There, Epstein likened the appeal to Christianity to the radical appeal to the ancient constitution. See James Epstein, "Some Organisational and Cultural Aspects of the Chartist Movement in Nottingham," in James Epstein and Dorothy Thompson (eds.), *The Chartist Experience: Studies in Working-Class Radicalism and Culture, 1830–1860* (London: Macmillan, 1982), 249–51.
[104] D.C.R.O., M/Du 34, Primitive Methodist Circuit Accounts, Durham, 1828–1836.
[105] Walker, "Growth of Wesleyan Methodism," 270–1.

that year, and within a fortnight further deaths were reported in Newcastle.[106] By the end of December, John Buddle, Lord Londonderry's colliery engineer, wrote that "the cholera is playing havock [*sic*] on the Tyne – at N[ew]castle, Gateshead, and most of the villages down the N[orth] side of the River to N[orth] Shields – Wallsend included. It broke out with great violence last Sunday, and on Monday I was informed there were 32 deaths and 59 new cases."[107] The dense colliery towns were particularly susceptible to the disease. In the last ten days of January 1832, twenty employees of Lambton Colliery died.[108] Henry Morton, never one noted for his sympathies toward the working class, expressed to Lord Durham the "harsh and unfeeling observation . . . that notwithstanding the mortality which has occurred, there still remains a larger population than is necessary – either as regards comfort to themselves or usefulness to society."[109]

Coming hard on the heels of the success of the union, the arrival of the cholera sparked a revival of Primitive Methodism.[110] Full membership in the Sunderland Circuit rose from 824 members in December 1831 to 916 in March 1832; those on trial jumped from 121 to 654. Total membership in the Sunderland, South Shields, and Newcastle circuits rose from 2,208 in 1831 to 3,360 in 1832.[111] Revival meetings garnered new converts in mining villages such as Walbottle and Newburn.[112] Between December 1831 and March 1832, Primitive Methodist membership at Hetton increased from 67 to 86, and the number of persons on trial for membership rocketed from 6 to 192. At Pittington, the site of Lord Londonderry's colliery, the combined total of full and trial members rose from 31 to 74; at Newbottle, Lord Durham's colliery, full and trial members increased from 9 to 22; and at Rainton, near another of Lord Londonderry's mines, full members rose from 10 to 72.

Furthermore, the incidence of the revival was centered not only in the mining towns but in the major cities as well – further evidence that the

---

[106] R. J. Morris, *Cholera 1832* (London: Croom Helm, 1976), 11, 60–4. See also the reports in the *Newcastle Chronicle*, 5 November 1831.

[107] D.C.R.O., Londonderry Papers, D/Lo/C 142(772), Buddle to Londonderry, 29 December 1831.

[108] Lambton Mss., Morton to Durham, 30 January 1832.

[109] Ibid., 16 January 1832.

[110] Colls, *Pitmen of the Northern Coalfield*, 151–4; however, this account of the effect of the cholera is seriously undermined by inconsistencies with regard to precise timing of unionization and the epidemic. Colls variously dates the cholera's crucial impact to July 1832, October 1831, and August 1832. See ibid., 97, 152, and 256.

[111] J. S. Werner, *The Primitive Methodist Connexion* (Madison: University of Wisconsin Press, 1984), 171.

[112] Morris, *Cholera 1832*, 145. On the links between Methodist theology and the cholera as God's retribution for sin, see ibid., 134–5.

revival was not linked specifically to trade unionism.[113] Three hundred new members were reportedly admitted to the local Methodist connexion after the cholera broke out in Gateshead.[114] Sunderland, which was especially hard hit, witnessed a similar growth in membership. In December 1831, there were 314 full and 82 trial members of Primitive Methodism. (Membership had already risen from the previous quarter when there were 294 full and 27 trial members.) By March 1832, the number of full-time members had risen to 377 and by the following quarter to 415. At the same time, trial members rose in Sunderland to a peak of 92 in March 1832.[115]

These numbers should make it clear that religious revivalism in Durham and Northumberland had an extremely complex and not necessarily direct connection to the working-class political movement. Thompson's suggestion that "religious revivalism took over just at the point when 'political' or temporal aspirations met with defeat" certainly cannot be supported by this evidence.[116] In the case of northeastern miners, religious revivalism followed in the wake of the workers' greatest success and came before their ultimate defeat. Moreover, rather than acting as an Halevyian siphon or Gilbertian safety valve for working-class aspirations, Methodism in Durham actually contributed its leadership to the formation of a radical trade union movement. Still, as has already been noted, this is not to say that religious enthusiasm and working-class militancy went hand in hand. The greatest growth of Primitive Methodist membership was not coeval with the union movement. The revival was incipient in some places at the beginning of the strike, but it was not widespread. Membership, trial and full, only took off after November 1831, and the proximate cause was not "political." It was caused by the sudden and inexplicable onslaught of the cholera.

One further point needs to be made. While trade unionism and evangelicalism did not share the same social foundations, their shared leadership meant that when the union was defeated, the revival sparked by the cholera died out as well. By June 1832, full and trial members of the connexion in the Sunderland Circuit had fallen by 266, and in the following quarter membership fell by a further 118 persons.[117] The circuit record keeper had little doubt as to the cause; in June 1832, it was noted, "decrease this quarter 266 on account of the pitman's stick." Why had the revival collapsed along with the union's defeat? The answer lies in both the nature of Methodist membership and the activity of the coal owners who mounted a concerted

---

[113] Evidence of Primitive Methodist membership refutes Morris's assertion that revivals linked to the cholera epidemic "did not take place in the big towns and cities." Ibid., 145.
[114] Ibid.
[115] D.C.R.O., M/Du 34, Primitive Methodist Circuit Accounts, Durham, 1828–1836.
[116] Thompson, *Making of the English Working Class*, 389.
[117] D.C.R.O., M/Du 34, Primitive Methodist Circuit Accounts, Durham, 1828–1836.

effort to drive the Methodist leaders from their company homes and to refuse to employ them. Morton, for example, refused to hire any of the pitmen's leaders after the 1831 strike, claiming that "the matter at issue is this, whether they or I shall have the control over the workmen of the colliery."[118] According to a nineteenth-century historian of the strike, Hepburn was refused employment in the collieries unless and until he foreswore all union activity.[119]

Deprived of a cadre of lay preachers, the revival could not maintain the momentum engendered by the cholera epidemic. The South Shields circuit record keeper believed that the loss of leaders combined with the disruptions caused by the strike contributed to the marked decline in membership. "It is the opinion of this Qr. Day Board," it was recorded, "that the loss of members is totally owing to the moovings [sic] which took place after the March Qr. Day and from the want of Leaders and Local Preachers."[120]

Therefore, the apparent contradition that the success of labor activism bore little relation to revivalism while the failure of trade unionism led directly to the collapse of the local evangelical movement can be explained through a fuller understanding of the role of Methodism in the mining community. That role reached beyond the boundaries of class meetings, Bible study, lay preaching, and evangelical revivals. In the 1830s, it encompassed a type of haven from the streams of traditional popular culture. When grievances weighed heavily on the miners, when working days were restricted by coal owners, or when fines were imposed punitively, these "serious" people were thrust forward to articulate demands, and not necessarily to provide spiritual leadership. As noted by a public defender of the miners, Robert Lowery, leadership was thrust upon these men of good character as if it were "a duty they owe their class."[121]

However, when the preachers' social position in the community was undermined by systematic intimidation, by an inability to find work, and by eviction from their homes, then the religious and political movements suffered alike. In the first half of the nineteenth century, the strength of Methodism lay in the sect's temporary and often short-lived ability to attract a literate and articulate minority of the northern working class. Furthermore, the significance of Methodism lay in the community's identification of these people as necessary and competent allies in their social struggles against the coal owners.

---

[118] David Spring, *The English Landed Estate in the Nineteenth Century: Its Administration* (Baltimore: Johns Hopkins University Press, 1963), 123.

[119] Fynes, *Miners of Northumberland and Durham*; J. L. Hammond and Barbara Hammond, *The Skilled Labourer* (1919; repr. London: Longman Group, 1979), 35.

[120] T.W.C.A., 1096/23, South Shields Circuit: Quarterly List of Members, 1832–1841.

[121] Harrison and Hollis, *Robert Lowery*, 75.

7

The transformation of market relations: Tommy Hepburn's union, 1831

The mere competition of producers, if left to the natural laws of distribution – free labor, entire use of its products, and voluntary exchanges – would be entirely of the exhilarating instead of the depressing species; and supported by increasing intelligence, would be ever on the advance with the increase of improvements in the arts, and of prudential and other moral habits in the people, till every laborer under equal security (casualties excepted) would be also a capitalist.

> William Thompson, *An Inquiry into the Principles
> of the Distribution of Wealth Most Conducive
> to Human Happiness* (1824)

The rise of a trade union in the early nineteenth century was an event of great moment to the entire working-class community. To later generations who grew up in the northern coalfield, strikes acted mnemonically to orient traditions and customs, justice and oppression, custom and change. To those who formed unions and who fought strikes, these were matters not only of wages and earnings, but of family, community, and, ultimately, class.

The ambiguous social position of the miner between that of a skilled craftsman and a common laborer provided part of the dynamic that made union movements difficult to mobilize and ideologically complex. Early-nineteenth-century trade unions in the northern coalfield embodied an implicit acceptance of the market economy, and union demands, based as they were upon the daily experience of labor, reflected the integral role that bargaining and negotiation played in the workplace.

As such, the union movements of the early nineteenth century were struggles to control the terms of market relations. Union power in the market ultimately rested on the defense of workplace bargaining, the control of the movement of labor, and restrictions upon output. Conversely, ownership and management sought control of the same product and labor markets through the maintenance of the "free" movement of labor as defined in the terms of the bond and the retention of the quota restrictions of the Vend. For trade unionists and capitalists alike, therefore, the resolution of the conflict between capital and labor was neither perceived nor acted upon as a conflict over the control of the labor process. Instead, the conflict was a struggle for market power.

Despite the relative insignificance of the struggle to control the labor process, this should not obscure the fact that the control of the market ultimately was a test of class power. In this sense, the struggle becomes evident in the realms of both the construction of rival political economies and alternative languages of market relations. In the end, in the northern coalfield, the struggle for market power was resolved by the use of financial and military force. Only then did the victors' class power define the market.

An analysis of the rise and fall of Tommy Hepburn's union provides a test for these assumptions. Hepburn's union eventually included over eleven thousand miners employed at more than fifty collieries spread over 350 square miles of northeast Durham and the southern portions of Northumberland. Remarkably, the union achieved its strike objectives and for a brief period forced the coal owners to accede to the union's demands. While the union's success was only short lived, a close reading of Hepburn's union movement can help to illuminate the nature of class struggle and the articulation of working-class ideology in the northern coalfield of the early nineteenth century.

The formation of Hepburn's union

On 1 January 1831, miners from the principal collieries on the Tyne and Wear met at Hetton-le-Hole, County Durham. According to the *Newcastle Chronicle*, the purpose of the meeting was to form a colliers' benefit society, and it was held with "the countenance and approval of the masters. It will in all probability," the report concluded, "become a very important society."[1] The *Chronicle* almost certainly overstated the extent to which the coal owners countenanced this meeting; no reference to it survives in the correspondence of the agents of the two great coal owners of the Wear, Lord Durham and Lord Londonderry. However, the paper did not exaggerate the importance of the meeting. Hetton-le-Hole became the center of Tommy Hepburn's union, and this New Year's Day meeting is likely to have marked the initial efforts to organize the movement.

Nearly two months later, on 26 February, the first mass rally of the miners was held on Black Fell near Gateshead. Reportedly, between eight and ten thousand people gathered on that Saturday to discuss their wages and the conditions of their employment.[2] According to the custom of local industrial

[1] *Newcastle Chronicle*, 15 January 1831.
[2] Ibid., 5 March 1831.

relations, petitions outlining the demands of the miners were to have been presented to the local colliery agents and viewers; however, no petitions promulgated at this time are extant in surviving colliery records.[3] On 6 March, over a week later, John Buddle was informed by his underviewer at Lord Londonderry's Rainton Colliery that the union had decided

> that they should not have a general Stop, in the first instance, but that "the men" of each Colliery should petition their Masters, at the ensuing Binding (20th Inst.) to restore the 1d./day taken off the Driver Boys last year and about 2 or 3 taken off the Putters, and that the day's Work of the Boys should be limited to 12 instead of 14 hours – the hewing prices to be settled by reference at each Colliery.[4]

The union leaders, believing that their petitions had gone unheeded, organized a second rally at Birtley Fell on Saturday, 12 March. The Birtley Fell rally resulted in the first set of petitions to the agents and viewers that have survived. At Cowpen Colliery, Northumberland, there were five demands: three requesting an increase in piece rates for hewers, one demanding a reduction of fines, and one desiring the reinstitution of an allowance for the purchase of candles.[5] Notably, this petition only requested changes in locally negotiated terms of payment. A second petition, however, soon followed that bore the stamp of regional trade unionism. On 15 March, John Watson, the agent at Cowpen, received a substantially longer petition that demanded not only further rate increases for putters and drivers but, more importantly, changes in the annual bond.[6] Similarly, at Walker Colliery, particular piece-rate demands were now joined by calls for changes in the bond. There, three demands for hewers' rate increases of up to eightpence per score were placed before the agent along with two further demands for increases for putters.[7] At Lord Londonderry's collieries, John Buddle promised to resist the demands of the pitmen. "The increase demanded by the Rainton, Pittington and Pensher pitmen," he wrote, "will amount to £10,000 a year – it *must be* resisted."[8]

Again, the point needs to be made that these piece-rate increases were peculiar to each colliery, and the union made no attempt to coordinate the hewers' demands. Indeed, it would have been foolhardy given the wide degree of variations in each colliery let alone across the northern coalfield. In 1831, while the Cowpen Colliery hewers asked for an eightpence per

---

[3] N.E.I.M.M.E., Bell Collection, vol. 11, 217, "Intended Meeting on the Town Moor."
[4] D.C.R.O., Londonderry Papers, D/Lo/C 142(658), Buddle to Londonderry, 6 March 1831.
[5] N.E.I.M.M.E., Watson Collection, Shelf 5/9/67; Shelf 5/9/73.
[6] Ibid.
[7] Ibid., Shelf 5/9/72.
[8] D.C.R.O., Londonderry Papers, D/Lo/C 142(661), Buddle to Londonderry, 12 March 1831.

score increase for double working, a threepence per yard increase for narrow work, and a twopence per score increase for wet working, the Walker Colliery hewers at the same time were demanding a sixpence per score increase for hewing Old Walls coal and separating large from small coal, an eightpence per score increase for hewing whole coal without separation, and a twopence per score increase for hewing Old Walls without separation.[9]

Hair's contention, therefore, that the uniformity of the bond worked against the freedom of the individual is only partly true at best.[10] That view perhaps entails a Whiggish assumption that the development of industrial relations ineluctably followed a path from individual action to collective bargaining. To be successful, union movements in the nineteenth-century coal industry needed to carefully balance the particularlity of individual collieries with the general demands of the region or district. Hepburn's union, as well as the United Colliers of 1825, did this by abdicating their influence over locally negotiated rates. This continued to be a principle of trade unionization in the entire period before the First World War despite the introduction of formal district-wide bargaining and the sliding scale.[11]

In 1831, the changes in the bond advocated by the union leadership and now added to local petitions were promulgated at Birtley Fell on 12 March: the limitation of child labor to twelve hours a day, alteration of the manner in which colliery workers could be laid idle without pay, and a separation of housing from the terms of employment.[12]

The first of these demands, the limitation of child labor, may in fact reveal the imprint of Methodism. While it has been argued that despite the appearance of a significant number of Primitive Methodist trade union leaders, including Tommy Hepburn, Methodist theology bore only a tenuous link to the trade union movement, the demands concerning child labor may be one of the few points of real contact.

Child labor, of course, was family labor. Most children were brought down the pits by their fathers or, as many bourgeois commentators were fond of noting, at the request of one of their parents.[13] The family, however, could

---

[9] N.E.I.M.M.E., Watson Collection, Shelf 5/9/67; Shelf 5/9/72.

[10] Hair, "Binding of the Pitmen," 1–2; see also Flinn, *History of the British Coal Industry*, 357–8.

[11] On the 1825 United Colliers Association, see Chapter 5, and *A Candid Appeal*, 9. Rowe clearly maintains that local bargaining continued to be the principal basis of wage rates in the northern coalfield up to the beginning of World War I. See Rowe, *Wages in the Coal Industry*, 49–51.

[12] N.E.I.M.M.E., Watson Collection, Shelf 5/9/67; N.E.I.M.M.E., Bell Collection, vol. 11, 217; Colls's claim that housing was not central to union activities before midcentury certainly is incorrect. See Colls, *Pitmen of the Northern Coalfield*, 263.

[13] *Reports from Commissioners: Children's Employment (Mines)*, P.P. (1842), vol. 16, 157, 618, 620, 621, 637, 677.

serve equally as a source of comfort and security as well as a source of exploitation and alienation. For every child, like John Otterson, whose father gave him a few candles so he would not have to sit in the dark for twelve hours or more,[14] there may have been another whose father "took him a clout" when he was "unruly, impertinent, or lazy."[15]

Both Thomas Laqueur and Robert Currie have shown that Methodism generally exhibited a special interest in the welfare and conduct of children.[16] Hugh Bourne, one of the founders of Primitive Methodism, claimed that while the innocence of children gave them "affinities with the Kingdom of God," it also made them very susceptible to the lure of sin.[17] Hence, the Primitives, like the Wesleyans, considered early education and religious training especially important. The demand for a twelve-hour working day may have grown out of this concern for children and their education. Tommy Hepburn plainly linked the reduction in working hours to education:

> Deprived of the privilege of light during the day, at night when their work is done, after a short time for refreshment, [the children] are put to bed, as their parents are anxious that they may get as much rest as possible previous to renewing their daily toil. No time is therefore left for mental recreation or for the improvement of their minds in any branch of knowledge: and considering they are put into the pit at the tender age of 6 or 7, and then bound apprentices to slavery till they arrive at the age of 21, no wonder that pitmen are brought up illiterate and ignorant of the most important and necessary branches of knowledge.[18]

Yet Hepburn's plea for the children of the pits, it should be noted, was not based on an evangelical fear of sin or the quest for redemption that one might expect from a Ranter preacher. He did not seek to exploit social problems for religious ends. Instead, Hepburn illuminated those secular concerns common to early-nineteenth-century working-class radicalism: improvement, knowledge, and literacy. They were concerns, nonetheless, that in the northern coalfield could separate an individual from the "rough and ready life" of their mates and encourage them to seek the company of like-minded friends in the Methodist societies.

The second of the three general demands of the union for changes in the bond was to separate housing from the terms of employment. The power of viewers and agents over their workers and over the labor market generally,

---

[14] Ibid., 163.
[15] *A Voice from the Coal Mines,* 25–6.
[16] Laqueur, *Religion and Respectability,* 9–18; Currie, *Methodism Divided,* 132–3.
[17] Kendall, *Origin and History,* vol. 1, 10, 48.
[18] N.E.I.M.M.E., Bell Collection, vol. 11, 218, "An Account of the Great Meeting of the Pitmen of the Tyne and Wear, on Newcastle Town Moor, on Monday, March 21, 1831."

as we have seen, was extraordinarily extended when it reached beyond the workplace and into the home.[19] Hepburn claimed that the viewers, power over "both bread and shelter" made the pitmen "slaves to the tempers and caprices of the Viewers."[20] Henry Morton, Lord Durham's agent, had no qualms about using his control of company housing to intimidate and control the pitmen. In the midst of Hepburn's strike, he wrote to Lord Durham:

> If I can get a magistrate tomorrow I mean to turn out 6 families at Newbottle – and if that does not produce the desired effect I shall proceed with 20 or 30 more immediately; there is nothing but decided and determined conduct that will bring these Brutes to reason.[21]

Most hewers certainly felt the steady pressure around binding time simply to renew their contracts rather than to pack up their homes and to search for a new job. After the Birtley Fell rally of 12 March, the union argued it was unfair that "if they are not bound again to the same colliery at which their time has just expired, and do not then immediately quit their houses, though they have no place to go, their furniture is, without further notice, put to the door."[22]

However, in these early stages of unionization, Hepburn and other union leaders failed to elaborate the manner in which they sought to separate housing from employment. In Hepburn's speech at Newcastle Town Moor on 21 March, he seemed to be groping toward a legal critique of the pitmen's tenure. "Being considered merely as tenants at will," Hepburn claimed, "on any difference with the Viewer, we are liable to be turned out at any time."[23] In 1832, he eventually did seek legal counsel with regard to the rights of the pitmen to their homes.[24] Local petitions at this time, however, reveal no clearly orchestrated demand. The petition from Cowpen Colliery simply protests "that article out of Bond [sic] which makes us Tenants at Will."[25] Yet at Walker Colliery, where the petition was presented by Benjamin Pyle, who was a leading activist in both Tommy Hepburn's union and later the national Miners' Association in the 1840s, the men demanded three months notice before quitting their homes.[26]

The third and final demand of the general union concerned the respon-

---

[19] See Chapter 3.
[20] N.E.I.M.M.E., Bell Collection, vol. 11, 218, "An Account of the Great Meeting of the Pitmen."
[21] Lambton Mss., Morton to Durham, 17 May 1831.
[22] N.E.I.M.M.E., Bell Collection, vol. 11, 218, "Intended Meeting on the Town Moor."
[23] Ibid., "An Account of the Great Meeting of the Pitmen."
[24] *Newcastle Chronicle,* 23 June 1832.
[25] N.E.I.M.M.E., Watson Collection, Shelf 5/9/73.
[26] Ibid., Shelf 5/9/72.

sibility of viewers to provide the miners a certain number of days' work each fortnight under the terms of the bond. Since the bond originated largely to secure skilled labor, the bonds by the 1830s stipulated that hewers would be guaranteed enough work to earn at least twenty-eight shillings a fortnight. However, if the pits had to be closed for repairs or maintenance, the hewers could be laid off without pay for a maximum of three consecutive days. On the fourth day, the hewers could either seek work elsewhere or receive lay-off pay at the rate of two shillings a day.[27]

The chronic complaint of Hepburn's union was that the owners manipulated these terms of the bond in order to control the product market. By laying the pits idle for three days, opening them for one, and then closing them for three more days, the coal owners' cartel could control production, force market prices up, and keep labor costs down. The first union broadside published after the Birtley Fell rally of 12 March specifically linked the dynamics of the cartelized industry to their own meager earnings:

> If [the men] are kept off work for three days, they can demand pay for the days after that, unless sent to work. To counteract this, the owners employ them the fourth day, and by that means can *cannily* keep them off the three days after without any further expense. In consequence of such conduct as this, the poor pitmen are often only employed four or five days in the fortnight.
>
> The reason why the owners give no better work is of course, from the combination among themselves, to send only a certain quantity of coals to market, in order to keep up prices.[28]

The union was absolutely correct, and under other circumstances, the owners admitted it. In 1829 and 1830, John Buddle testified before the Parliamentary Commission on the Coal Trade. When a member of the Commons' Committee asked him if wages were as good as they had been in 1810, Buddle responded that wage rates were comparable, but earnings were down: "The wages would have been quite as good [as in 1810] if we had employment for the people, but they are at very reduced work."[29] Estimating that hewers were "not making more than eight shillings or ten shillings a week,"[30] Buddle told the Lords' Committee that "we are not enabled in some places to give them more than three days [of work] a week"[31] – a conclusion that explicitly bears out the claims of Hepburn's union.

[27] Ibid., Bell Collection, vol. 11, 217, 218; *Newcastle Chronicle*, 26 March 1831; *A Candid Appeal*, 6–10; Flinn, *History*, 352–3; Colls, *Pitmen of the Northern Coalfield*, 45–51; Hair, "Binding of the Pitmen," 1–13; Scott, "The Miners' Bond," pt. 1, 70–1.

[28] N.E.I.M.M.E., Bell Collection, vol. 11, 217, "Intended Meeting on the Town Moor."

[29] *Select Committee into the State of the Coal Trade*, P.P. (1830), vol. 8, 315.

[30] *Select Committee of the House of Lords on the State of the Coal Trade*, P.P. (1830), vol. 8, 274.

[31] Ibid., 66–7.

Privately, Henry Morton admitted to Lord Durham that their hewers earned only 13s. 6d. a week during 1830–1, a shilling less than the earnings "guaranteed" each fortnight by the bond.[32]

When the colliers met at Birtley Fell on 12 March, the representatives of the owners gathered at the Coal Trade Office in Newcastle. There, under the chairmanship of R. W. Brandling, the cartel resolved to resist almost all increases in wages and to turn the terms of the bond further against the miners. They decided that trappers' and putters' rates were not to be increased and children's working hours were to remain at fourteen hours per day. Hewers' rate increases were to be resisted "but the Managers of each Colliery [were] to exercise discretion in making such advances as particular cases may require." No binding money was to be given that would serve to attract workers to sign the bond, and all fines were to remain unchanged. Finally, hewers were to be guaranteed nine working days a fortnight or enough work to earn twenty-eight shillings, unless a pit "shall be rendered unfit for working," and binding time was to be moved back to the first Saturday after 21 February.[33]

The last decision concerning binding time was an explosive issue, and the attempt to move it back into February is indicative not only of the owners' willingness to risk widespread labor unrest, but also of their recognition of the continuing struggle over control of the labor market. In 1809, an earlier attempt to shift binding time from its traditional period in October to January led to a industry-wide strike and to the establishment of 5 April as the current binding date.[34] The owners' object, of course, was to situate binding time so as to limit the mobility of labor as much as possible and restrict the competition for hewers that drove up costs. The owners' decision at their subsequent meeting of 28 March 1831 to move the binding date further back to 1 January was an even bolder move in this direction.[35]

The miners, for their part, certainly were aware that this would significantly weaken their bargaining position, as testified by the 1809–10 strike. In 1831, John Watson, the Cowpen Colliery agent, noted that "many of the men think the alteration [of binding time] to Jany. is doing them no good & only throwing them into Winter for travelling round the Collys."[36]

Before the bonds were published on 19 March, allowing a fortnight before their actual signing on 5 April, the union had already planned a third rally

---

[32] Lambton Mss., Morton to Durham, 16 March 1831.

[33] N.R.O., General Meetings of the Coalowners of the Rivers Tyne & Wear, 81–2, 88.

[34] Ashton and Sykes, *Coal Industry*, 95–6; Hair, "Binding of the Pitmen," 1–13; Flinn, *History*, 354–5; Colls, *Pitmen of the Northern Coalfield*, 81–2.

[35] N.R.O., General Meetings of the Coal Owners of the Rivers Tyne & Wear, 28 March 1831, 83–4.

[36] N.E.I.M.M.E., Watson Collection, Shelf 5/9/76c.

to be held on Newcastle Town Moor. On Monday, 21 March, an estimated 24,000 supporters assembled there. Throughout the morning, the miners marched peacefully through the town, each colliery organized in ranks of four or five abreast. "A few took refreshment in town," a local handbill read, "but most of them proceeded directly to the place of meeting."[37] The miners from Hetton Colliery were among the last to join the rally, and they dramatically entered the grounds between noon and one o'clock. In all, miners from forty-seven collieries from the Tyne, Wear, and Blyth were represented.[38]

The principal speaker was Tommy Hepburn, the Ranter preacher and hewer from Hetton Colliery. He reiterated the demands of the union and propounded a policy of strict moderation. Above all, Hepburn hoped to create a union based upon peace, order, and publicity. He described the miners' movement as an ethical struggle that would bring public pressure, in part through the medium of the press, to bear upon the coal owners and viewers. "It was only by good order and peaceable demeanor," he said that day, that "they could hope to obtain the approbation of the magistracy and the good will of the community at large."[39]

To this end, he deplored machine breaking and violence, and promised to turn over to the magistrates anyone who violated the peace. Hepburn, moreover, was acutely aware that the union was a fragile and vulnerable instrument. Subject to intimidation and threats of force from the owners and the magistracy as well as from dissension within the movement, Hepburn constantly reiterated the values of publicity and mutuality. His speech at Newcastle was typical of many of his later speeches urging "every man to take care of and protect his neighbour." "Let the pitmen be faithful to one another," he said, "and while they deported themselves peaceably they may hope to obtain a redress of their grievances, and experience the sympathies and support of the community at large."[40]

Support for this policy of moderation seems to have been deep seated. By all accounts, the structure of the union was decidedly democratic, and union leaders were held fully and directly accountable. The union was based

---

[37] Ibid., Bell Collection, vol. 11, 218, "An Account of the Great Meeting of the Pitmen."

[38] *Newcastle Chronicle*, 26 March 1831; N.E.I.M.M.E., Bell Collection, vol. 11, 218, "An Account of the Great Meeting of the Pitmen."

[39] N.E.I.M.M.E., Bell Collection, vol. 11, 218, "An Account of the Great Meeting of the Pitmen." Hepburn's faith in publicity and the power of the press bears a startling resemblance to the opening paragraphs of Thomas Hodgskin's *Labour Defended Against the Claims of Capital* (1825; repr. London: Labour Publishing, 1922), a resemblance, I argue, that may not have been accidental.

[40] N.E.I.M.M.E., Bell Collection, vol. 11, 218, "An Account of the Great Meeting of the Pitmen."

upon local committees at each colliery that organized the popular selection of three delegates, or deputies, to represent the colliery at union meetings.[41] Regular union meetings were held in public houses, most frequently the Sign of the Cock in Newcastle, and attended by between 150 and 200 delegates.[42] At least in the early stages of the union movement, the temper of the meetings was decidedly proper and civil. Buddle recounted to Lord Londonderry the observations of two viewers, Thomas Foster and George Hunter, who were sent to meet the pitmen in April 1831:

> They found the Delegates assembled to the amt. of about 200 all seated at tables so contrived as to bring them all into one large room. Hebburn [sic] was chairman, and Dixon (of Cowpen) was secretary – pen, ink, and paper was placed at the corners of all the tables. When Hunter and Foster were introduced by a Backworth delegate, Hebburn was on his legs speaking. After repremanding [sic] the delegate, for having introduced "the Viewers" so unceremoniously without first duly announcing them, and stating the object of their visit – they were asked what they came there for?[43]

The placement of pen, ink and paper at the corners of all tables might be taken as a tangible symbol of the direct democracy embodied in the union. Delegates were liable to be recalled at any time. Hepburn himself reminded the miners of this when the union's defeat was near.[44] The delegates themselves frequently understood their role as representatives of the pitmen in a literal sense. No crucial actions were taken by the delegate committee without the general consent of the rank and file. At a meeting between the delegates and viewers in the early days of the strike, the delegates plainly "declared that they were not authorized by the Body of the Pitmen to agree finally to the propositions of the Viewers, and that they could only communicate and consult with the Body upon them."[45] The mass rallies that punctuated the union movement thus became open forums of popular sovereignty where these matters were discussed and decided upon.

The miners, nonetheless, were not political reformers. Throughout the

---

[41] D.C.R.O., Londonderry Papers, D/Lo/C 142(658), Buddle to Londonderry, 6 March 1831; Lambton Mss., Morton to Durham, 10 March 1831.

[42] N.R.O., General Meetings of the Coal Owners of the Rivers Tyne & Wear, 9 April 1831, 85–6; D.C.R.O., Londonderry Papers, D/Lo/C 142(672), Buddle to Londonderry, 10 April 1831.

[43] D.C.R.O., Londonderry Papers, D/Lo/C 142(672), Buddle to Londonderry, 10 April 1831, also quoted in Colls, *Pitmen of the Northern Coalfield*, 255.

[44] *Newcastle Courant*, 2 June 1832.

[45] N.R.O., General Meetings of the Coal Owners of the Rivers Tyne & Wear, 29 April 1831, 93–4.

movement, the union repeated its loyalty to the king and obedience to the laws of the land. The Reform Bill movement barely affected the union despite the fact that Lord Grey and Lord Durham were at the center of that controversy. Hepburn spoke only once at a rally of the middle-class Northern Political Union during which time he supported the extension of the franchise to £5 householders, a reform that would have excluded the miners, in the hope that their representatives would help break the power of monopolies.[46] In all the reported public rallies of the pitmen, the union leaders rarely expressed a desire to seek the reform of Parliament or an expansion of the franchise. Indeed, at one of the last rallies of Hepburn's union, it was generally resolved "that in order to display our loyalty to the King and his government, we will stand or fall by adhering to the laws of this realm."[47]

Why was this? To a large extent, the nonpartisan stance of the union grew out of its tactics and its ideology. In order to obtain the "approbation of the magistracy and the good will of the community," the union was forced to adopt the terms of a discourse that divorced it from partisan political controversy. The miners' struggle was thus portrayed first as a matter of ethics, not politics. The initial public formulation of the miners' demands sought justice and fairness in their dealings with their employers. One of the miners' strike songs, "The Pitman's Complaint," pointedly called upon the bourgeoisie to express the same compassion for the miners that they showed toward West Indian slaves: "The Indian slaves for freedom groan / We have a greater cause to moan / You often pity slaves abroad / But we have now a greater load."[48] Hepburn, it may be recalled, referred to child laborers as "bound apprentices to slavery."[49]

Religious ideals were similarly freely appropriated in a way that may be construed so as to avoid overt political partisanship. Old Testament homilies blended a feeling of righteousness with that of resistance to tyranny. John Johnson, one of Hepburn's lieutenants, frequently followed this tack and once came close to calling Hepburn a modern-day Moses. Recounting the brief interregnum of Hepburn's success, Johnson said:

> Before their late victory, pitmen were circumstanced exactly like the Israelites under the hard-hearted Pharaohs of old, and laboured under Egyptian darkness. But the Lord did for them, as he had done for the children of Israel; he had made a way for their escape. The enemy, however, still pursued and

---

[46] Peter Cadogan, *Early Radical Newcastle* (Consett: Sagitarrius Press, 1975), 66.
[47] *Newcastle Courant,* 2 June 1832.
[48] N.E.I.M.M.E., Bell Collection, vol. 11, 228.
[49] Ibid., 218, "An Account of the Great Meeting of the Pitmen."

pressed upon them, and would certainly defeat their object if they were not steadfastly united to each other.[50]

And the discourse of child labor and slavery was combined once again by Charles Parkinson, a Primitive Methodist union delegate who proclaimed that the miners "were now struggling not only for the emancipation of themselves, but of their posterity."[51]

More importantly, however, religion and child labor became mobilizing metaphors for Hepburn's union because they articulated an ideology that principally sought to redefine market relations.[52] Recourse was made to themes of tyranny and slavery principally because they reflected the dysfunction of the marketplace. In a handbill distributed after the initial success of Hepburn's union, the pitmen of Gateshead Park Colliery appropriated both the discourse of radicalism and religious resistance to oppression in order to express their newfound power in the market:

> Our Employers themselves have sent a most malicious Report afloat concerning us, saying, that we want to be Masters, and make them our HIRED SLAVES; this we absolutely deny; for, thank God! although they call us Dupes and ignorant Men, we have learned what it is to be Servants, and to conduct ourselves with Propriety and Respect towards them. We will submit to be Servants, but not BOND SLAVES, for our Forefathers and Ourselves have been too long subject to their Bondage, Tyranny, and Oppression; we will henceforward put our Trust in Him, who sees all Things and judgeth all Things according to their Merits or Demerits, and hope to live in better Days than our Forefathers; for having last year got Part of the Burden taken off our Shoulders, and in Consequence of which, have rather tasted the Sweets of Liberty, we are, therefore, determined to be no more entangled with the Yoke of Bondage.[53]

The content of this discourse, however, did not reflect the vision of a disintegrating moral economy. One should not be misled into believing that Hepburn's union, as it has been claimed for other unions of this period, reflected little more than "the desirability of re-establishing a world governed by customary expectations and just agreements regulating the conduct of masters and men."[54] As we have seen, in their economic relations with their

---

[50] *Newcastle Courant*, 21 April 1832.

[51] *Newcastle Chronicle*, 23 June 1832.

[52] This bears some similarity to Robert Gray's analysis "The languages of factory reform in Britain. c. 1830–1860," in Joyce, *Historical Meanings of Work*, 143–79.

[53] N.E.I.M.M.E., Bell Collection, vol. 11, 413, "To the Public from the Late Pitmen of Gateshead Park Colliery."

[54] This is Gareth Stedman Jones's general assessment of trade union ideology in the late 1820s and early 1830s, in "Rethinking Chartism," *Languages of Class*, 114.

masters the pitmen were not, as it were, trapped by the dead weight of custom. On the contrary, their behavior was penetrated by the acceptance of market relations.

Rather, the political ideology of Hepburn's union bore a greater degree of similarity to some of the ideas propagated by those who have come to be known, very inadequately, as the Ricardian socialists.[55] No tangible evidence exists that either Hepburn or his lieutenants read works such as Thomas Hodgskin's *Labour Defended Against the Claims of Capital* (1825) or William Thompson's response entitled *Labour Rewarded: The Claims of Labour and Capital Conciliated* (1827), but there are surprising echoes of these works in the elaboration of the union's ideology.

First, Hepburn's union sought neither the destruction of capitalism nor the dispossession of capitalists, an approach implicit in Hodgskin's *Labour Defended*.[56] Unlike Owenite socialist thought during this era, for example, union manifestoes made no attempt to distinguish between productive and unproductive classes.[57] This may have been because the "aristocracy of coal" was not perceived as either idle or indolent. Yet without this dis-

---

[55] Two recent works dealing with the Ricardian socialists are Noel W. Thompson, *The People's Science: The Popular Political Economy of Exploitation and Crisis, 1816–34* (Cambridge University Press, 1984), and Gregory Claeys, *Machinery, Money and the Millenium: From Moral Economy to Socialism, 1815–1860* (Princeton: Princeton University Press, 1987). Both Thompson and Claeys reject the explanatory value and empirical validity of the term "Ricardian socialist" to describe the varied contributions of these writers. Thompson prefers "Smithian socialism" to emphasize these writers' rejection or unfamiliarity with Ricardian theory as opposed to their more profound acquaintance with Smith's work. Claeys, however, prefers to carve up the traditional categorization of these economists and substitute a new order of "Owenite socialists" that emphasizes their affiliation with Robert Owen's work. Although the evidence presented here would tend to support Thompson's claims, I have eschewed both new labels in order to facilitate the identification and recognition of the historical location of these writers.

As opposed to the debate over popular political economy, John Belchem's work in particular has emphasized working-class radicalism's dedication to "constitutional mass pressure from without for the restoration of the constitutional democratic rights of all," especially after the Spa Fields meetings of 1816–17. As noted, Hepburn's movement bore little resemblance to these movements of popular constitutionalism. See John Belchem, *'Orator' Hunt: Henry Hunt and Working-Class Radicalism* (Oxford: Oxford University Press, 1985), 4; idem, "Republicanism, Popular Constitutionalism and the Radical Platform in Early Nineteenth-Century England," *Social History*, 6 (1981), 1–32.

[56] Hodgskin, *Labour Defended*, 83–92; Thompson, *People's Science*, 98–9, 102–3. Hodgskin admitted to a difference between useful, productive "masters" and idle "capitalists." But without acknowledging a structural antagonism between capital and labor, he was left to argue that the sole solution to oppression and exploitation was a more equitable distribution of "wages" among laborers and masters.

[57] Claeys, *Machinery, Money and the Millennium*, 135–6.

tinction, it proved impossible to construct an analysis of society in which labor was the sole source of wealth, and the miners never adopted this theory of value.

Second, Hepburn's union sought the establishment of a set of market relations that was deemed to be both fair and equitable. Like Thompson's *Labour Rewarded*, or more significantly his earlier work, *An Inquiry Into the Principles of the Distribution of Wealth Most Conducive to Human Happiness* (1824), Hepburn's union embodied the argument that the source of labor's exploitation lay in inequitable exchange relations. However, neither Hepburn nor his union ever linked a theory of inequitable exchange relations to a labor theory of value as did Robert Owen and some of his followers.[58] Instead, like some of the work of both Thompson and Hodgskin, Hepburn apparently believed that equality and freedom from exploitation would result from a truly competitive economy from which the elements of force, fraud, and monopoly had been eliminated.[59] Trade unionism thus became, to use Noel Thompson's words, a way to "redress the disparate bargaining strength of labourers and capitalists and thus contribute to a more perfect functioning of the market."[60]

To the trade unions of the northern coalfield, the marketplace was an acceptable forum for industrial relations provided that the structure of market relations was fair and equitable. The founders of Hepburn's union were neither naive nor disingenuous when they described their goal as to "see the day when the poor shall have fair play as well as the rich."[61] Nor were the United Colliers of 1825 insincere in the exercise of their "unalienable right of nature, to demand equity, and 'fair play', from all our fellow men, with whom we are connected."[62] They were expressing a belief in equitable exchange relations within the structures of early industrial capitalism.

However, "fair play" did not (and could not) connote the retention of customary wages or traditional work practices.[63] Instead, it expressed a demand for equal and reciprocal power over capital and labor. Tommy

---

[58] On Owenite theory of just exchanges, see ibid., 138–42; Thompson, *People's Science*, 102.

[59] Thompson, *People's Science*, 98–9; Claeys, *Machinery, Money and the Millennium*, 91–9. Claeys notes, however, that William Thompson's endorsement of the benefits of free competition became much less fulsome in *Labor Rewarded* than it had been in *An Inquiry into the Principles of the Distribution of Wealth Most Conducive to Human Happiness* (1824), which was published three years earlier.

[60] Thompson, *People's Science*, 98.

[61] N.E.I.M.M.E., Bell Collection, vol. 11, 217.

[62] *A Voice from the Coal Mines*, 33.

[63] This definition was briefly suggested by Stedman Jones in an earlier version of "Rethinking Chartism" called "The Language of Chartism" in Epstein and Thompson, *The Chartist Experience*, 19.

Hepburn, for example, told a rally of the miners in 1832 that "they wanted an agreement [with the owners], but, at the same time, they wanted one-half of that agreement to be of their own making."[64] Robert Arkle, a leader of Hepburn's union, explained that reciprocal power in the marketplace was the key to the union's success: The "owners had matured their system . . . by unanimity; [and] the men must, therefore, be unanimous in supporting theirs. . . . The owners had showed the men the way to combine, and they might thank themselves if the men had done them any harm."[65]

This perception of industrial relations as based on theories of equitable exchange did not preclude, however, the identification of an irresolvable conflict between capital and labor. Equitable exchange theories were not mere bourgeois illusions, as Marx later argued.[66] When James Losh, a local attorney and coal owner, sought to explain to the pitmen that there existed an identity of interests between labor and capital, "An Old Pit-man," a prominent union publicist, responded by quoting none other than Adam Smith to the effect that wages were based upon contracts in the formation of which the interests of employers and laborers were "by no means the same."[67] On the one hand, this lends support to Noel Thompson's claim that the recognition of the intrinsic antagonism of capital and labor was an argument that constituted one of the unique contributions of the Ricardian socialists, and one, moreover, that was achieved through their engagement with Adam Smith as opposed to David Ricardo.[68] On the other hand, it elucidates the complex manner in which union movements came to elaborate class-based criticisms of society and the varied sources of that critique.[69]

Hepburn's union did not yearn for a stable, customary past or even the revivification of a system of mutual obligations.[70] Embedded in the language of morality, justice, and fairness, even in the language of master and servant, was a desire for reciprocal but independent market power. Significantly, as we have seen, Tommy Hepburn hoped that the reform of Parliament would

---

[64] *Newcastle Chronicle*, 23 June 1832.
[65] Ibid.
[66] J. E. King, "Utopian or Scientific? A Reconsideration of the Ricardian Socialists," *History of Political Economy*, 15: no. 3 (1983), 346–7.
[67] *Newcastle Chronicle*, 18 February 1832. Adam Smith's argument precedes his famous condemnation of masters' combinations to influence the movement of wages in bk. 1, chap. 8, *Wealth of Nations*, 74.
[68] Thompson, *People's Science*, chap. 4.
[69] This is contrary to Gareth Stedman Jones's argument that unions of this period did not express class-based criticisms of society; see "Rethinking Chartism," in *Languages of Labour*, 117.
[70] On the rhetoric of mutual obligations, see Gray, "Languages of factory reform," 152–3.

lead directly to the advent of free trade and the destruction of monopolies.[71] Benjamin Pyle, a leader of both Hepburn's union and the later Miners' Association, argued that the miners sought neither to dispossess the masters nor to usurp their authority over industry. But they demanded the power to deal equitably with them:

> They wanted liberty of conscience; they sought to obtain the freedom of thought common to every Englishman – to have such rights and privileges that were conformable to their situation as servants – and an equal right with their employers to make the best bargain they could for their labour.[72]

The construction of new market relations

The history of the strike itself reveals the extent to which Hepburn's union was wedded to a vision of equitable market relations, a vision that could be achieved only by a significant redistribution of power in the market. To this end, the union first sought to break the coal owners' monopoly power over the market for labor principally through a negotiated restructuring of the bond. However, during the course of the strike, that vision, which combined both equity and arbitration, had to contend with rank-and-file demands for employment and a return to work in the face of the appearance of blackleg laborers. Under these pressures, the union leadership reluctantly adopted a policy of output restriction and thus extended its control into the sphere of production. Concurrently, the union took formal control of the labor market by licensing the movement of labor.

On 5 April 1831, miners at collieries throughout Durham and Northumberland ceased work and removed their gear from the pits. Since the Newcastle rally, the bargaining positions of both owners and the union had evolved and hardened. By the start of the strike, the owners expressed a willingness to reduce the length of children's work days to twelve hours and to eliminate truck in the collieries. (Truck, the payment of workers in goods rather than money, had never been prevalent in the Northeast, but some viewers and agents did sell candles, powder, and other provisions to the miners often at inflated prices. As Mitchell has noted, the owners thus lost nothing by the elimination of truck.)[73] They refused to yield, however, on guarantees of both the number of work days per fortnight and the earnings for that period. In addition, the owners rescinded their earlier offer of an advance on drivers' wages and now sought to move binding time further back into the year to

---

[71] Cadogan, *Early Radical Newcastle*, 66.
[72] *Newcastle Courant*, 21 April 1832.
[73] *Newcastle Chronicle*, 26 March 1831; Mitchell, *Economic Development*, 237.

the first Saturday in January. Finally, the owners agreed among themselves to lock out all colliers who refused to sign the bonds on these terms.[74]

By this time, the pitmen's grievances had also changed significantly. Along with the three original demands of revised working guarantees, a twelve-hour working day for children, and the separation of housing from the terms of employment, the union now struck out at the foundations of managerial authority by seeking the reduction of fines for poor work and the right to have the corves measured regularly.[75] Still, the union continued to exhibit a willingness to bargain and negotiate. On 6 April, they notified the coal owners' cartel that delegates from the union were willing to meet with a delegation of viewers "to discuss all matters in dispute," a meeting that was then scheduled for 11 April.[76]

The next day, however, the union received an implicit endorsement from a very unlikely source. Eleven Northumberland magistrates offered to mediate the dispute and arranged a meeting of representatives of the two sides in Newcastle.[77] The mayor of Newcastle, Archibald Reed, had been in close contact with the union well before the strike began and may have exerted some influence upon the magistrates. As early as 21 March, the mayor had attended the Newcastle union rally and similarly offered to serve as a mediator between the miners and their employers.[78] At first, some viewers responded equivocally. Upon first hearing of the news, John Buddle thought, "It can do no harm."[79] However, Henry Morton complained to Lord Durham that the magistrates had no right to interfere in a dispute solely between masters and men,[80] a line of argument that shortly convinced Buddle as well. Within a matter of days, Buddle was claiming to Lord Londonderry that the magistrates had "given the pitmen fresh courage and vigour under the impression that the coal-owners were *cribling* (crouching to them) and that they have applied to the magistrates to negotiate terms for them with the pitmen!!!"[81]

It appears unlikely that such efforts at mediation were the result of the

---

[74] N.R.O., General Meetings of the Coal Owners of the Rivers Tyne & Wear, 28 March 1831, 83–4; D.C.R.O., Londonderry Papers, D/Lo/C 142(668), Buddle to Londonderry, 29 March 1831.

[75] N.E.I.M.M.E., Bell Collection, vol. 11, 288; *Tyne Mercury*, 12 April 1831.

[76] N.R.O., General Meetings of the Coal Owners of the Rivers Tyne & Wear, 9 April 1831, 85–6.

[77] *Newcastle Chronicle*, 9 April 1831.

[78] N.E.I.M.M.E., Bell Collection, vol. 11, 218, "An Account of the Great Meeting of the Pitmen." See also ibid., vol. 11, 288.

[79] D.C.R.O., Londonderry Papers, D/Lo/C 142(671), Buddle to Londonderry, 8 April 1831.

[80] Lambton Mss., Morton to Durham, 10 April 1831.

[81] D.C.R.O., Londonderry Papers, D/Lo/C 142(672), Buddle to Londonderry, 10 April 1831.

magistrates' sole desire to stabilize local social relations.[82] While this may
have been the effect of such actions, it seems more likely that at least in this
instance, the magistrates acted out of fear, self-interest, and weakness. At
the commencement of the strike, the Home Office had advised the magistrates that the infantry and dragoons in the area were undermanned and
would be useless in the case of a major disturbance.[83] The local commanding
officer had told the magistrates directly that it was "more than ever necessary
to throw those Gentlemen upon their own resources, and that they may
protect their property with a civil instead of a military force." Moreover, he
insisted that troops would be called out only in the instance of a large-scale
riot but not to protect individual properties.[84]

Still, Buddle was essentially correct: The union was indeed bolstered by
the magistrates' action. At a preliminary meeting between two viewers and
the union delegates on 9 April, the union resolved "not to meet with the
Viewers alone, but would meet the Viewers and Coalowners in [the] presence
of the Magistrates." However, the union also made it clear that the mediation
of the magistrates should not be taken as an indication of the miners' acceptance of their arbitration as binding. "The Magistrates [were] to determine who produced the best Arguments," the union representatives explained, "but they would not allow the Magistrates either to mediate or
arbitrate for them, or to decide any thing between the Parties."[85] Given the
vast experience that the miners had with bargaining and bargaining institutions at the local level, this sophisticated and pragmatic approach to the
mediation of the magistrates should not be surprising.

The formal meeting between the miners and the owners' representatives
was held on 11 April at the Moot Hall in Newcastle. Almost as soon as the
talks began in earnest, the owners' unwillingness to negotiate became evident.
The first topic of discussion was the number of guaranteed working days
per fortnight. The union offered to accept the relevant clauses of the bond
if they were amended to guarantee eleven instead of nine working days each
fortnight. The owners' representatives refused outright, and the meeting
broke up shortly thereafter "under some marks of disapprobation on the
part of the Pitmen."[86] Archibald Reed was convinced that the owners had

[82] McCord, "The Government of Tyneside," 28–30; Colls, *Pitmen of the Northern Coalfield*, 250–1; also see Chapter 4.
[83] P.R.O., H.O., 40/29/40, Col. Sir Hew Ross to Maj. Gen. Bouverie, 6 April 1831.
[84] P.R.O., H.O., 40/29/34, Col. Ross to Maj. Gen. Bouverie, 4 April 1831.
[85] N.R.O., General Meetings of the Coal Owners of the Rivers Tyne & Wear, 9 April 1831, 85–6; D.C.R.O., Londonderry Papers, D/Lo/C 142(672), Buddle to Londonderry, 10 April 1831.
[86] N.R.O., General Meetings of the Coal Owners of the Rivers Tyne & Wear, 11 April 1831, 87.

no desire to come to terms with the miners. A week earlier, he had related the following to the local troop commander:

> So far from the coal owners being anxious to come to an immediate accommodation with their workmen, the very reverse is the case, they consider for their interest to enhance the price of coals by a temporary stand, and for that reason the terms they have offered to the pitmen are such as to be very beneficial to themselves if accepted, and if rejected (which they will be) they are well content.[87]

The owners' report of the Moot Hall meeting underscores several important aspects of the form and structure of the miners' union.[88] First, the owners' negotiators were fairly certain that the nearly two hundred local delegates had deputed seven men to negotiate for them. However, it was also apparent to them that these seven delegates had not been authorized to conclude a settlement. (The same, it should be noted, was the case for the owners' representatives.) Sovereignty rested in the rank and file, and they would not yield that sovereignty either to outside mediators or to their own representatives.

Second, the union leaders were under significant pressure from the rank and file. The delegates, it was reported, "were frequently interrupted and broke in upon by others." These interruptions apparently became so disruptive that the magistrates warned the miners against a breach of the peace. The seven special delegates clearly walked a very thin line between their role as negotiators and as deputies of the rank and file. Buddle claimed that "great jealousy exists amongst the body, against the delegates, and the seeds of disunion are sown." He advised Lord Londonderry at this time to remain quiet and aloof and the union would fall apart from its own internal dissension.[89] This dialectic between delegated authority and rank-and-file sovereignty was one of the most important factors in determining the form and policy of Hepburn's union.

Despite the failure of these negotiations, encouraging signs of an impending settlement appeared in the days following the Moot Hall meeting. The union now circulated a set of resolutions that dropped all of their previous demands except those regarding the number of guaranteed days per fortnight and the alteration of the limitations of how and when corves could be remeasured.[90] The owners, of course, had already agreed to the

---

[87] P.R.O., H.O., 40/29/34. Col. Ross to Maj. Gen. Bouverie, 4 April 1831.

[88] N.R.O., General Meetings of the Coal Owners of the Rivers Tyne & Wear, 11 April 1831, 87.

[89] D.C.R.O., Londonderry Papers, D/Lo/C 142(678), Buddle to Londonderry, 14 April 1831.

[90] N.R.O., General Meetings of the Coal Owners of the Rivers Tyne & Wear, 11 April 1831, 90; D.C.R.O., Londonderry Papers, D/Lo/C 142(676), 13 April 1831.

elimination of truck and the limitation of children's workdays to twelve hours. Buddle hoped that the strike's end was imminent even though the prices of work at individual collieries had yet to be negotiated.[91]

Even at this point, the manner in which the union demands were formulated reflected their explicit acceptance of market relations and a belief in the market's ultimate efficacy. Indeed, union resolutions reveal the desire to unbridle the productive capacity of individual hewers and to let the market determine how this may affect their earnings. Thus the miners resolved on 11 April that "we do not agree to be laid idle two days in the fortnight. And that we take our chance for Employment as it may happen to occur, waiving the [guarantee of] 28s. per fortnight."[92] The Tyneside miners, who held a separate rally on 14 April, expressed the same sentiments in a slightly different manner. They resolved that "the Men and Boys are willing to abide the Risk of the Fluctuations of the Coal Trade, provided they are bound to work eleven days in the fortnight."[93] The miners believed that the coal cartel had manipulated the mechanics of the market to limit the supply of coal and that this was the principal cause of their low earnings.[94] If these restraints were removed, labor would be unshackled and both capital and labor would suffer or profit alike from the vicissitudes of the commodity market.[95]

That brief glimmer of hope and accommodation quickly faded. The third week of April witnessed an escalated attack by the coal owners upon the union. On 16 April, a violent antiunion tract was published in the *Durham Chronicle*, and on 19 April the *Tyne Mercury* repeated its outspoken condemnation of the miners.[96] During the same week, several collieries tried to start up work with nonunion labor. Union members were called on to stop blacklegs at Jesmond, Ouston, Hetton, Bedlington, Beamish, Tanfield, Shilbottle, and Netherton collieries.[97] The week of agitation ended with a mass rally of the pitmen at Jarrow on the banks of the Tyne.

Henry Morton reported to Lord Durham that for the first time in the course of the strike, religion was becoming mixed in a dangerous way with the union movement. Colliery meetings, he said, now began with prayers

---

[91] D.C.R.O., Londonderry Papers, D/Lo/C 142(675), Buddle to Londonderry, 13 April 1831.
[92] N.R.O., General Meetings of the Coal Owners of the Rivers Tyne & Wear, 13 April 1831, 90.
[93] N.E.I.M.M.E., Bell Collection, vol. 11, 226.
[94] Ibid., 217.
[95] Correspondingly, Colls's argument that miners' unions from the 1820s were characterized by efforts to control the autonomy of workers cannot be supported without significant qualification. Colls, *Pitmen of the Northern Coalfield*, 29.
[96] N.E.I.M.M.E., Bell Collection, vol. 11, 282, "The Pitmen of the Tyne & Wear"; ibid., 295.
[97] D.C.R.O., Londonderry Papers, D/Lo/C 142(678), Buddle to Londonderry, 14 April 1831; D/Lo/C 142(682), 16 April 1831; D/Lo/C 142(689), 20 April 1831; N.E.I.M.M.E., Bell Collection, vol. 11; 295, 297; Lambton Mss., Morton to Durham, 21 April 1831.

"invoking the almighty to strengthen and prosper the cause."[98] As early as 27 March, Morton had been aware that Primitive Methodists were leading the union at Lord Durham's collieries, and upon first meeting them he admitted that "their acuteness and intelligence surprised me."[99] Morton soon became obsessed with the influence of the Ranter preachers upon the trade union movement to a far greater degree than most of his colleagues. Morton thought the union was the result of the religious fanaticism of a few "designing individuals" who were able to exert their influence over the "exceedingly ignorant" masses.[100] However, others, especially John Buddle, certainly never conflated the union with evangelicalism. Only once, as has been noted, did he condemn one union leader, Charles Parkinson, for being a "slick headed *ranting* knave."[101] Matthias Dunn, who often dealt directly with Parkinson, never made note of any particular evangelical motivation among the colliers at Hetton even though that colliery was at the epicenter of the union movement.[102] The success of evangelical religion among the miners was not due to its ability to sanctify a resistance movement that was born of oppression and exploitation. Evangelicals led the union movement because they possessed the skills necessary to organize a union and to present the miners' case to both the owners and the public.

The first break in the strike came during the last week of April and the first week of May 1831. On 23 April, the union again offered to send seven delegates to meet with the same number of representatives of the coal owners.[103] It appears that economic pressures were beginning to weigh heavily, and both sides were eager to reopen negotiations. About a week earlier, the union had distributed the last of the strike pay it controlled. This amounted to between twenty-six and twenty-eight shillings per man, enough to last about a fortnight.[104] The coal owners also showed some signs of desperation. On 27 April, Lord Durham became the first major coal owner to start up production again after he forced open Lumley Colliery under the protection of forty uniformed soldiers.[105] Similar efforts were shortly

[98] Lambton Mss., Morton to Durham, 16 April 1831.
[99] Ibid., 27 March 1831; 6 April 1831.
[100] P.R.O., H.O., 40/29/89, Morton to Maj. Gen. Bouverie, 8 June 1831.
[101] D.C.R.O., Londonderry Papers, D/Lo/C 142(851), Buddle to Londonderry, 21 June 1832.
[102] N.C.L., Matthias Dunn's Diary, 1 December 1831.
[103] N.R.O., General Meetings of the Coal Owners of the Rivers Tyne & Wear, 25 April 1831, 92; ibid., Committee Meetings of the Coal Owners of the Rivers Tyne & Wear, 23 April 1831, 320; D.C.R.O., Londonderry Papers, D/Lo/C 142(692), Buddle to Londonderry, 23 April 1831.
[104] Lambton Mss., Morton to Durham, 16 April 1831; D.C.R.O., Londonderry Papers, D/Lo/C 142(688), Buddle to Londonderry, 19 April 1831.
[105] Lambton Mss., Morton to Durham, 27 April 1831.

underway at Hetton Colliery, while Lord Londonderry took the extraordinary step of meeting personally with his pitmen to entice them back to work.[106]

This next meeting of delegates and viewers was held at the Turk's Head Inn in Newcastle on 29 April. Negotiations were fruitful, honest, and constructive. Tentative agreements were reached on binding time (to remain at the fortnight before 5 April), housing (the viewers agreed to give the men a fortnight's notice before eviction), and the adjustment of the corves (the viewers agreed to allow "the privilege of seeing that the Corves are not larger than is necessary to deliver 20 Pecks at [the] Bank"). There was a slight disagreement over the precise definition of twelve working hours for the boys and day workers. The delegates argued that working hours should be counted from the time at which the last boy reaches the bottom of the pit, while the viewers argued that twelve hours should be counted from the time work actually started.

Two significant obstacles remained however: the number of guaranteed days of work, or guaranteed earnings, each fortnight and the system of fines. In the former instance, the men wanted to be guaranteed eleven days' work for twenty-five fortnights each year and to be provided enough work to earn thirty-three shillings a fortnight. The viewers offered to guarantee them ten days' work a fortnight but would not promise any minimum level of earnings. With regard to fines, the miners' argued that they should only be liable to forfeit the price of each corve that was deemed unsatisfactory by management. The viewers were adamant in their report to the coal owners that they "cannot concede this." Since the viewers had little control at the point of production, their managerial control of the quality of output rested on their ability to levy punitive fines of up to a shilling for "foul corves." These heavy fines were the bulwark of managerial authority and could not be bargained away.[107]

The 29 April meeting ended with an agreement to meet again on 3 May. The following meeting, however, was tense and acrimonious. A new demand for an increase in putters' rates that had first been tabled on 29 April led to a lengthy and inconclusive debate. Some movement was noticeable over the question of working hours, and the owners further sought to accommodate the union by specifying that the corves could be measured without prior notice and that management would be allowed not more than three days to adjust the corves. The number of working days again proved to be a major stumbling block even though both sides had revised their positions

---

[106] D.C.R.O., Londonderry Papers, D/Lo/C 142(700), Buddle to Londonderry, 8 May 1831; N.E.I.M.M.E., Bell Collection, vol. 11, 295.

[107] N.R.O., General Meetings of the Coal Owners of the Rivers Tyne & Wear, 29 April 1831, 93–4; N.E.I.M.M.E., Watson Papers, Shelf 5/9/76a; ibid., Bell Collection, vol. 11, 327.

since the previous meeting. The union moved to accept the owners' offer of ten guaranteed working days each fortnight but lowered their demand for minimum earnings to thirty shillings. The owners, for their part, now agreed to a minimum level of earnings, but it was only twenty-eight shillings a fortnight. This the union "positively refused."[108]

The 3 May meeting revealed more than the union's continued dedication to collective bargaining. The viewers reported that the union delegates now had been given full authority by the rank and file to reach a settlement with the owners, something they had not had on 29 April. Of equal importance was the fact that the seven union delegates were no longer wholly united under Hepburn's leadership. Hepburn appeared to the viewers "to be labouring under great mental agitation." Moreover, "it was evident that he did not possess the same degree of authority over his colleagues as at the former Meeting."[109]

The crisis of Hepburn's leadership reflected an even greater crisis within the union movement. Hepburn's attempt to build a respectable trade union movement based upon the principles of equitable exchange was being undermined by popular violence and direct rank-and-file action. "Stripping," a form of northern charivari, surfaced for the first time. William Garth, a wasteman at Mount Moor Colliery who refused to join the union, was dunked in a pond, covered with dung, and stripped naked except for his shoes and socks.[110] (At the Newcastle Town Moor rally of 21 March, Hepburn had specifically urged the miners not to strip anyone.)[111] At Jesmond Colliery, there was a case of machine breaking in which the strikers also threw burning hay down the pit.[112] Rumors spread that the union was planning to institute a *taxation populaire*. Buddle heard "talk of fixing a maximum price of corn, butcher's meat, etc." as well as an increase of harvesting wages.[113] The hostile *Tyne Mercury* reported cases of wandering pitmen extorting "contributions" from local farmers.[114] The viewer's house at Cowpen Colliery was broken into and food and drink taken. The next night, the viewer received a letter that protested social inequality and demanded justice:

> I was at yor hoose last neet, and myed mysel very comfortable. Ye hey nee family, and yor just won man on the colliery, I see ye hev a greet lot of rooms, and big cellars, and plenty wine and beer in them, which I got ma share on.

[108] N.R.O., General Meetings of the Coal Owners of the Rivers Tyne & Wear, 3 May 1831, 96–7.
[109] Ibid.
[110] *Newcastle Chronicle*, 30 July 1831, "Report of the Summer Assizes."
[111] N.E.I.M.M.E., Bell Collection, vol. 11, 218.
[112] Ibid., 297.
[113] D.C.R.O., Londonderry Papers, D/Lo/C 142(699), Buddle to Londonderry, 4 May 1831.
[114] N.E.I.M.M.E., Bell Collection, vol. 11, 297.

Noo I naw seom at wor colliery that has three or fower lads and lasses, and
they live in won room not half as gude as yor cellar. I don't pretend to naw
very much, but I naw there shudnt be that much difference. The only place
we can gan to o the week ends is the yel hoose and hev a pint. I dinna pretend
to be a profit, but i naw this, and lots of ma marrows na's te, that wer not tret
as we owt to be, and a great filosopher says, to get noledge is to naw wer
ignerent. But weve just begun to find that oot, and ye maisters and owners
may luk oot, for yor not gan to get se much o yor awn way, wer gan to hev
some o wors now. I divent tell ye ma nyem, but I was one of yor unwelcome
visitors last neet.[115]

This bubbling of spontaneous popular activism began to subside when
Lord Londonderry's resistance to the union weakened noticeably, and consequently, the prospects of the union brightened. On 5 May, Lord Londonderry, acting as the only ennobled magistrate in the region, helped to
disperse a local miners' rally held on Black Fell. As the men left the rally,
Tommy Hepburn fell in with the marquess and told him that if the owners
agreed to guarantee the men thirty shillings a fortnight the strike would be
called off immediately.[116] Londonderry suggested that "he would endeavour
to induce the coal owners to accede to this demand" and encouraged the
union to send several delegates to meet with the owners at the Coal Trade
Office in Newcastle the following day.[117]

The cartel responded positively to Londonderry's initiative and sent
three viewers, John Watson, Thomas Easton, and Nicholas Wood, to
meet the delegates on 6 May. The cartel, with Lord Londonderry in the
chair, hoped that the parties could still reach "an amicable agreement on
this one point." But this was not to be. Despite Hepburn's suggestion of
the previous day, the other delegates were determined to tie the question
of punitive fines to that of guaranteed earnings, the "cloven foot" as
Henry Morton called it.[118] Subsequently, the owners resolved to open
their collieries under the protection of civil and military forces and to
force the dissolution of the union.[119]

Londonderry then took several extraordinary steps to come to an agreement with the colliers at his mines.[120] On 7 May, he and Lady Londonderry

[115] Fynes, *Miners of Northumberland and Durham.* The letter is reprinted in Thompson, *Making of the English Working Class*, 715.
[116] N.R.O., General Meetings of the Coal Owners of the Rivers Tyne & Wear, 6 May 1831, 100; N.E.I.M.M.E., Bell Collection, vol. 11, 327.
[117] N.E.I.M.M.E., Bell Collection, vol. 11, 327.
[118] Lambton Mss., Morton to Durham, 6 May 1831.
[119] N.R.O., General Meetings of the Coal Owners of the Rivers Tyne & Wear, 6 May 1831, 100–101; D.C.R.O., Londonderry Papers, D/Lo/C 142(698), Buddle to Londonderry, 4 May 1831.
[120] Colls, *Pitmen of the Northern Coalfield*, 90–1.

met personally with the strikers.[121] The following day, Londonderry circulated an open letter to his pitmen in which he guaranteed them thirty shillings a fortnight if they would leave the question of fines to be settled on his honor and by the good sense of his viewers.[122] Further, he authorized Buddle to give in to the advances in piece rates demanded by the pitmen. Buddle estimated that this would amount to an additional charge of £3,000 per annum, provided that the men started work immediately and left the union.[123]

Londonderry's actions had a startling effect. Buddle was "extremely mortified and vexed."[124] Londonderry's letter to the pitmen, Morton wrote to Lord Durham, "is most absurd and ridiculous – and the construction put upon it by the public, is, that he is either at a stand for money, or taken leave of his wits."[125] Indeed Londonderry's actions are open to a variety of interpretations. Colls has argued that he broke ranks with the cartel because his financial position was in a state of near chaos and he needed to start up production without delay.[126] The building of Seaham Harbor, at this time, certainly was a constant drain on his resources, and it was nearly complete.[127] By the first of June, Londonderry's account with one bank was overdrawn by £22,000 in order to meet current expenses while Seaham expenditures averaged over £29,000 per annum in 1830–1.[128]

However, this was not the only problem faced by Lord Londonderry. At the same time, his son, the young Lord Castlereagh, was being opposed at the Irish elections in Down, and the marquess was compelled to make a "hurried and forced departure for Ireland."[129] He thus felt that he had to conclude the strike of his pitmen in Durham prior to his return to his Irish estates. In addition, Londonderry may have felt no reasonable obligation to resist the demands of the miners. On 4 May, the day before Londonderry had met Tommy Hepburn at Black Fell, the cartel, with Londonderry in the chair, had decided to break off collective bargaining sessions with the union. Moreover, they had resolved that all owners were to stand by their last collective bargaining offer with regard to fines, working days, houses, and putting rates, and that local piece-rate prices should remain at the level

---

[121] D.C.R.O., Londonderry Papers, D/Lo/C 142(700), Buddle to Londonderry, 8 May 1831.
[122] N.E.I.M.M.E., Bell Collection, vol. 11, 300, 327.
[123] D.C.R.O., Londonderry Papers, D/Lo/C 142(702), Buddle to Londonderry, 9 May 1831.
[124] Lambton Mss., Morton to Durham, 9 May 1831.
[125] Ibid., 7 May 1831.
[126] Colls, *Pitmen of the Northern Coalfield*, 90.
[127] D.C.R.O., Londonderry Papers, D/Lo/C 142(704), Buddle to Londonderry, 12 May 1831.
[128] Ibid., D/Lo/C 142(718), Buddle to Londonderry, 2 June 1831; Sturgess, *Aristocrat In Business*, 72–4.
[129] N.E.I.M.M.E., Bell Collection, vol. 11, 300, 327.

originally offered to the men at binding time in March. Yet the cartel's final
resolution of 4 May was not unambiguous. While collective bargaining sessions were deemed a "total failure," the coal owners and viewers were
nonetheless allowed to continue to negotiate with the miners at each colliery,
albeit limited by the cartel's resolutions.[130] It was this last defense that
Londonderry employed in a public letter to the coal trade published at the
end of May.[131]

In any event, Londonderry's actions signaled the impending breakup of
the coal owners' resistance. The men at Londonderry's Rainton, Pittington,
and Pensher collieries remained firmly united to the principles of regional
trade unionism and told Buddle that they could not accept the marquess's
offer without permission from the central delegates.[132] However, the miners
at two of the largest collieries on the Wear, Durham's Newbottle Colliery
and Hetton Colliery, agreed to sign the bonds apparently based on offers
of thirty shillings each fortnight, comparable to Hepburn's informal offer to
Lord Londonderry on 5 May.[133]

The union delegates hastily gathered in Newcastle. They were now faced
with the problem of whether to allow some miners to return to work under
more or less favorable terms or to keep the entire union out on strike until
they obtained a regional settlement. Fearing that the movement was splin-
tering, the delegates declared that no miner should sign a bond until all
collieries on both rivers acceded to the union's demands.[134]

The miners at the Londonderry collieries, however, rejected the union
leadership. Their actions reveal the complex relationship between the regional demands of the union over the terms of the bond and the immediate sphere of industrial relations that concerned bargaining over piece
rates. Lured by Buddle's offer of advances in hewers' piece rates and in
drivers' day rates above those offered to the men at Newbottle and Hetton collieries, the Londonderry pitmen signed their bonds on the night of
12 May.[135] Hearing of these terms, the miners at Newbottle and Hetton
quit work again and demanded the same piece-rate advances from their
viewers.[136] At the same time, the cartel attempted to regroup after Londonderry's defection. They resolved to match Londonderry's offer of

---

[130] N.R.O., General Meetings of the Coal Owners of the Rivers Tyne & Wear, 4 May 1831,
   98–9.
[131] N.E.I.M.M.E., Bell Collection, vol. 11, 327.
[132] D.C.R.O., Londonderry Papers, D/Lo/C 142(702), Buddle to Londonderry, 9 May 1831.
[133] Ibid.; Lambton Mss., Morton to Durham, 14 May 1831.
[134] D.C.R.O., Londonderry Papers, D/Lo/C 142(705), Buddle to Londonderry, 12 May 1831.
[135] Ibid., D/Lo/C 142(706), Buddle to Londonderry, 13 May 1831.
[136] Ibid., D/Lo/C 142(707), Buddle to Londonderry, 14 May 1831.

thirty shillings a fortnight but refused to countenance the piece- and day-rate advances.[137]

The resumption of work at Londonderry's three large collieries as well as the brief reopening of Newbottle and Hetton obviously began to undermine the unity of the cartel. Since the strike had shut down the mines in April, London coal prices had risen by over 20 percent.[138] The actions of the largest Wear coal owners allowed them to take full advantage of the price rise while the smaller Tyneside owners continued to bear the costs of their idle collieries. The Tyneside coal owners either could maintain their support of the cartel and watch the Wear collieries reap the fruit of their resistance or they could break ranks and begin to work their own mines.

By 18 May, the latter course had been adopted. Collieries throughout the coalfield initiated a concerted effort to evict union families from their colliery-owned houses and to hire nonunion labor in their stead. Evictions took place at Friar's Goose, Percy Main, and Hebburn collieries as well as at Lord Durham's Newbottle Colliery and at Hetton.[139] This, however, had an unintended effect: The union spirit that seemed to have been lagging suddenly began to revive. The coal owners' oppression that had previously been defined largely in the language of inequality and exploitation was now felt in very real terms. One union handbill explained:

> While the pitmen, by a general union, have endeavoured to obtain a redress of their grievances, which have been fairly stated ... and are quietly appealing, merely by *the Force of Argument*, to the candour and good sense of their employers and the community at large; the coal owners, armed with Civil and Military power, appear wishful to appeal only to *the Argument of Force*.[140]

More importantly, the union leadership adopted new strike tactics forced upon them by the rank and file. At the few working collieries, the union members instituted production restrictions and began to contribute a portion of their earnings to the strike fund. At Lord Londonderry's collieries, the miners contributed a quarter of their earnings to the union.[141] At Lord Durham's collieries, the workers limited their earnings to a maximum of

---

[137] N.R.O., General Meetings of the Coal Owners of the Rivers Tyne & Wear, 11 May 1831, 102.

[138] *Report of the Commissioners appointed to enquire into the Several Matters relating to Coal in the United Kingdom*, P.P. (1871), vol. 18, 208.

[139] N.E.I.M.M.E., Bell Collection, vol. 11, 280; Lambton Mss., Morton to Durham, 17 May 1831.

[140] N.E.I.M.M.E., Bell Collection, vol. 11, 280.

[141] D.C.R.O., Londonderry Papers, D/Lo/C 142(710), Buddle to Londonderry, 18 May 1831; D/Lo/C 142(711), Buddle to Londonderry, 19 May 1831.

three shillings a day, forcing Morton to give work to more men.[142] That this
restriction was the result of rank-and-file action is confirmed by the announcement of 18 May by the union delegates: "It appears to be the intention
of the pitmen of the collieries where they are bound, to support their brethren
of the unbound collieries, to the amount of one half of their earnings, if
required, while the latter are as determined as ever to stand off until their
prices...shall be granted."[143] Albeit reluctantly, the union leadership was
forced to accept as policy the spontaneous actions of their members. This
was, moreover, a policy that would shortly lead to success.

The combination of new union tactics and the rivalry among the coal
owners broke the cartel's resistance. On 20 May, Jonathan Brandling gave
in to the union demands at South Shields Colliery. The following day, the
coal owners' committee meeting in Newcastle weakly tried to hold the cartel
together by recommending that no advance in wages be made without their
approval.[144] It was to no avail. On 23 May, the cartel informally acknowledged their defeat by giving the owners full discretion to reach agreements
with their men as they deemed appropriate.[145] By 28 May, twenty-eight
further collieries on the Tyne had yielded to the union.[146] By 1 June, only
ten collieries on both the Tyne and Wear were not working, and they gave
in within another month.[147]

The union now controlled production throughout the region. The spon-
taneous system of mutuality that had been created during the crisis of mid-
May became established as the new mode of working. Section VIII of the
Rules and Regulations of the Coal Miners' Friendly Society promulgated on 4
June 1831 limited hewers' earnings to four shillings per day clear of fines.[148]
Fortnightly contributions were levied, and as Colls has shown, membership
was broadened to include putters, drivers, and, later, lower-level managerial
personnel.[149] Furthermore, the union now took control of the labor market
and prohibited the movement of workers without the possession of a license
granted by the union's central committee. The dialectical development of
the movement was complete. The union had evolved from an organization

---

[142] Lambton Mss., Morton to Durham, 17 May 1831.
[143] N.E.I.M.M.E., Bell Collection, vol. 11, 280.
[144] N.R.O., General Meetings of the Coal Owners of the Rivers Tyne & Wear, 21 May 1831,
106.
[145] N.R.O., Committee Meetings of the Coal Owners of the Rivers Tyne & Wear, 23 May
1831, 321.
[146] *Newcastle Chronicle*, 28 May 1831.
[147] N.R.O., General Meetings of the Coal Owners of the Rivers Tyne & Wear, 1 June 1831,
107.
[148] N.E.I.M.M.E., Bell Collection, vol. 11, 312–13.
[149] Colls, *Pitmen of the Northern Coalfield*, 90, 249.

dedicated to freeing market forces into the cartel's antithesis: a workers' movement that controlled production and the movement of labor. The trade union had temporarily succeeded in their struggle for market power.

The strike movement of 1831 culminated in a victory rally on Boldon Fell on 13 August. The previous week, the delegates had announced that the rally was meant to express the union's patriotism, peacefulness, and obedience to the laws of the land.[150] Throughout the morning of the thirteenth, the miners walked in procession behind the banners of their collieries. It was estimated at the time that by midday between eight and twelve thousand people had assembled. Five of the delegates addressed the rally and emphasized the necessity of acting peacefully and legally. Tommy Hepburn warned that the miners "were to be cautious and not step too far – not take a rash step which would lose them the good opinion of the public, subject them to the operation of the laws, and make them more enemies than friends." Robert Arkle urged the pitmen not to "forfeit the protection of laws, and the good-will of the public" by engaging in any misconduct. Charles Parkinson, Buddle's "slick headed ranting knave," suggested that the miners endeavor to educate their children in both civil and religious matters, but also to remember that "they were but servants, and if they assumed a power which did not belong them, they would ovethrow the system which they had set up, and do away all the good which their recent exertions had procured for them."

At the conclusion of the meeting, Hepburn suggested that the miners should send an address of good will and loyalty to the king, an announcement that was supported by a show of hands. The rally then disbanded after colliery bands played "God Save the King" and three cheers were given for the king. The following Thursday, the delegates dispatched the union's address of loyalty to the crown signed by 11,561 miners from fifty-seven collieries in Durham and Northumberland.[151]

This rally epitomized the legalistic and formal veneer of respectability that the union leadership cultivated and that was supported by the rank and file. It could hardly be argued, as some have done, that the union's success reflected a rapprochement between radicalism and evangelical fundamentalism.[152] The expression of the union's ideology often trod along the well-worn paths of resistance to oppression and exploitation expressed equally in the languages of faith and reason. But the union's ultimate success in 1831 was predicated upon rank-and-file activism and their leaders' conception of their democratic responsibility to reflect and articulate these concerns. Toward

---

[150] *Newcastle Chronicle,* 6 August 1831.
[151] Ibid., 20 August 1831; 27 August 1831; *Newcastle Courant,* 20 August 1831.
[152] Colls, *Pitmen of the Northern Coalfield,* 189–202.

this end, the leading cadre of lay preachers eschewed much of their fundamentalist rhetoric except where it reflected the general demands for justice and fair play. Moreover, the working out of these demands was decisively transformed during the strike movement. Once concerned with freeing the market for labor and commodities from the deadening hand of the cartel, the union eventually established a control over labor that was the reciprocal of the owners' control of capital. Market relations had been transformed by the success of Hepburn's union, but they had not been abrogated.

8

Epilogue: class struggle and market power

The failure of Hepburn's union

The control of both production and the movement of labor exercised by
Tommy Hepburn's union was generally uncontested between May 1831 and
April 1832.[1] The struggles that resumed after that date revealed the full
extent to which the contest for power in the market was also understood as
a class struggle. The immediate economic interests of capital became of
considerably less significance than its ultimate control of labor. The bitter
memories of 1832 were the result of this transformation in which power in
the marketplace became nothing less than a cipher for class power.

Sporadic wildcat strikes over particular issues at individual collieries continued during the period after Hepburn's initial victory. Such was the case
at Lord Durham's Newbottle Colliery, where a strike occurred in response
to Henry Morton's refusal to employ union delegates.[2] Most union-

---

[1] Hepburn and his union usually merit a place among the pantheon of labor heroes. However,
many of the extant sources, particularly local newspapers and Home Office papers, are skewed
to more fully document the events of 1832 rather than those of the previous year. This led
earlier historians, particularly the Hammonds, to portray a far more accurate picture of the
collapse of the union rather than its genesis. The works of Fynes and the Hammonds on
this second phase of the strike are still unsurpassed in many respects. See Hammond and
Hammond, *The Skilled Labourer*, and Fynes, *Miners of Northumberland and Durham*. Less
satisfactory accounts appear in Edward Welbourne, *The Miners' Unions of Northumberland and
Durham* (Cambridge University Press, 1923), and Sidney Webb, *The Story of the Durham
Miners, 1662–1921* (London: Fabian, 1921). It is particularly unfortunate that the authors
of the "official" history of the British coal industry sponsored by the old National Coal Board
relied on Welbourne's account of the strike. This led them to claim at one point that Hepburn's
union failed in 1831 after a two-week strike and then failed again after several weeks in 1832.
See Church, *History*, 677. Overall, Colls's account in *Pitmen of the Northern Coalfield*, 93–7,
247–54, combines a vast knowledge of local sources with an almost breathless presentation.
[2] Lambton Mss., Morton to Durham, 1 June 1831. Similarly, Lord Londonderry's Pittington
colliers struck when they heard rumors that neighboring Black Boy Colliery had raised
trappers' wages by twopence a day. When the rumor proved to be false, the strike was called
off. See D.C.R.O., Londonderry Papers, D/Lo/C 142(751), Buddle to Londonderry, 7 July
1831; D/Lo/C 142(758), Buddle to Londonderry, 11 July 1831.

sanctioned strikes during this period, however, were limited to the prevention and intimidation of blackleg labor.[3]

In at least two cases, collieries were able to break the union and hire nonunion workers,[4] while at most other collieries the employment of black-legs elicited customary forms of rough justice from the local rank and file, sometimes to the distress of the more conservative union leadership. In February 1832, for example, four union miners were arrested for assaulting nonunion lead miners at Gosforth Colliery; that same month, nonunion pit-men were also attacked at Cramlington Colliery.[5] The following month, five teenaged putters attacked and stripped a nonunion hewer at Hetton Colliery. Later that same evening, union delegates visited the victim and apologized. They returned his clothes and gave him a half-crown for his troubles.[6]

Overall, however, the union's control was surprisingly secure. Buddle noted that "at present, we have a *strict regulation* of quantity, and will continue to have, unless something occurs to abate the spirit of domination which the pitmen are exercising. It cannot be denied that they have *entire dominion* over the Trade."[7] Of course, the owners did not suffer in silence. The language of master and servant as well as that of political economy was deployed against the union. R. W. Brandling, the chairman of the coal own-ers' cartel, complained that the pitmen were exercising "an improper Con-troul over both Masters and Servants."[8] The *Tyne Mercury*, a Whig paper, argued:

> The question at issue between the pitmen of the Tyne and Wear and coal-owners is now plainly who shall be masters? All persons who have workmen under them ... must see that there is an end to business altogether if servants are to rule over their masters, to tell them how much work is to be done, and who are to do it.[9]

Other antiunion advocates, such as the pamphleteer "Scrutator," employed the language of political economy to register their reaction to the union. Blithely unaware of the irony of defending the cartel on such terms, Scrutator argued, nonetheless, that "labour as commodities must be regulated by the proportion of supply and demand." The four shilling limit was "absurdly

---

[3] The "Waldridge Outrage," for example, when over a thousand miners assembled at Waldridge Colliery and stopped the pumping engines there in order to prevent the introduction of Weardale lead miners, seems to have had the union's imprimatur. *Newcastle Chronicle*, 10 March 1832.

[4] Colls, *Pitmen of the Northern Coalfield*, 93.

[5] *Newcastle Courant*, 11 February 1832.

[6] *Newcastle Chronicle*, 7 April 1832.

[7] D.C.R.O., Londonderry Papers, D/Lo/C 142(718), Buddle to Londonderry, 2 June 1831.

[8] *Newcastle Chronicle*, 8 September 1832.

[9] *Tyne Mercury*, 12 June 1832.

capricious" and a dangerous "assumption of power and [the] right of dictation."[10]

The union argued in return that, unlike the cartel, theirs was a moral union for the improvement of the working class. Hepburn's famous call for the creation of "itinerating libraries" based upon union subscriptions was part of the union's ethical rationale.[11] At other times, it was suggested that the four shilling limit was more akin to a goal than an outright restriction. "It takes the *very best men*, to work with 'extraordinary exertion' from six to *eight* hours to earn four shillings," a union advocate claimed. He added: "In my own case, I know I have made about four shillings a-day since May, in eight or nine hours, but I know many collieries where the hewers have not made more than three shillings in eleven and twelve hours."[12]

Perhaps there was one significant reason why capitalists did not immediately attempt to take direct action against output restriction, if this could have been done. Ironically, restriction served the same purpose as the coal owners' cartel. Buddle repeatedly claimed that the miners had succeeded in regulating total output precisely where the coal owners had failed. In May 1831, he wrote to Lord Londonderry that "this Regulation will soon become a more strict one than the Coal-owners could ever effect."[13] At times Buddle certainly wished he could increase the collieries' output to cover some extraordinary expenditures. Nevertheless, he recognized that while "nothing but *Hemp*" would unite coal owners, at least the goal of regulating production was being achieved by the union.[14] Lord Durham, for one, made astounding profits during the strike. His colliery profits totaled more than £34,500 in 1831.[15] It was no wonder that Lord Durham's agent explained, "I cannot brook the idea of being dictated to by these rascals – tho' I am unwilling to gratify these feelings by the diminution of your Profits."[16]

Nevertheless, the industrial magnates' actions against the union in 1832 underscore the proposition that the union movement was recognized as a struggle for market power that itself resolved into a class struggle. As Lord Londonderry was advised at the end of May 1832: "*It is said*, that [the pitmen] complain of too hard labour – small Wages – working in bad air and above all of the *tyranny* of the Coal-owners!!! They totally forget the

---

[10] Scrutator, *An Impartial Enquiry into the Existing Causes of Dispute between the Coal Owners of the Wear and Tyne and their late Pitmen* (Houghton-le-Spring, 1832).
[11] *Newcastle Chronicle*, 21 April 1832.
[12] Ibid., 24 March 1832.
[13] D.C.R.O., Londonderry Papers, D/Lo/C 142(715), Buddle to Londonderry, 29 May 1831.
[14] Ibid., D/Lo/C 142(714), Buddle to Londonderry, 28 May 1831; D/Lo/C 142(713), Buddle to Londonderry, 26 May 1831.
[15] Lambton Mss., Morton to Durham, 1 March 1832.
[16] Ibid., 1 June 1831.

tyranny exercised by themselves."[17] Despite evidence that profits were sustained during the time in which the union controlled the labor and product markets, the coal owners marshaled the financial strength of local banks and the military strength of the state to defeat the union. Given the choice between profit and power, the coal owners' actions unequivocally proclaim that their social power and its corollary, the control of the market, was more important than their immediate economic interests. The recognition of the true nature of the struggle, along with the effects of the cholera epidemic of 1831–2, precipitated the collapse of Hepburn's union movement.[18]

The cholera, which struck Sunderland at the end of October 1831, quickly raged through the dense colliery villages.[19] At Hetton, Matthias Dunn recorded in his diary in January 1832 that "nearly *100* men off work from illness at present – chiefly Cholera Morbis."[20] Again in February he noted, "Cholera again breaking out with great violence. 3 deaths last night. 75 hewers off work by it."[21] Within a week, there were over thirty deaths at Wallsend in December 1831[22] and twenty deaths at Lambton Colliery at the end of January.[23]

Providing sickness and death benefits chiefly for cholera victims drained substantial resources from the union's funds. Between May 1831 and June 1832, the union paid out over £13,000 in such benefits.[24] In March 1832, the delegates claimed that they had distributed over £700 at Hetton alone.[25] Thus, when binding time recommenced in March and April 1832, the union was significantly less prosperous than it otherwise could have been.

Meanwhile the coal owners began to grope to restore a semblance of order among themselves almost as soon as the strike ended in 1831. On 31 August, the coal trade recommended that all proprietors should refuse to hire workmen without a complete enquiry into their employment history, a measure designed to wrest some control over the labor market from the union.[26] In

---

[17] D.C.R.O., Londonderry Papers, D/Lo/C 142(831), Buddle to Londonderry, 30 May 1832.
[18] See this author's "The State, Capital and Workers' Control during the Industrial Revolution," *Journal of Social History*, 21, no. 4 (1988), 717–34.
[19] See Morris, *Cholera 1832*.
[20] N.C.L., Matthias Dunn's Diary, 25 January 1832.
[21] Ibid., 9 February 1832.
[22] D.C.R.O., Londonderry Papers, D/Lo/C 142(772), Buddle to Londonderry, 29 December 1831.
[23] Lambton Mss., Morton to Durham, 30 January 1832.
[24] N.R.O., 2FO/1/10, "An Account of the Receipts and Expenditures of the Collieries belonging to the Pitmen's Union, commencing May 27, 1831, to and with June 23, 1832." See also Hammond and Hammond, *Skilled Labourer*, 30–1.
[25] *Newcastle Chronicle*, 10 March 1832.
[26] N.R.O., Committee Meetings of the Coal Owners of the Rivers Tyne & Wear, 31 August 1831, 330.

mid-September, the coal owners formed a strike indemnity fund both to reimburse firms that had accommodated the military during the strike and to assist collieries that still were being laid idle by the union.[27] Within several months, the cartel had distributed over £1,500 to three collieries.[28] In January 1832, the coal owners agreed to a further indemnity plan whereby collieries would be reimbursed at the rate of five shillings for each chaldron that they fell short of their quota.[29]

In March 1832, the coal owners' formulated the principals and tactics on which to base their counterattack upon the miners' union. They generally ignored the union's control of production, still further evidence of the owners' acceptance of their unwillingness to wrest control of the labor process, and concentrated on breaking the union's authority over the movement and employment of labor. In that month, the coal owners' committee published a brief report on the state of the coal trade intended to reach an audience of coal consumers. They attempted to argue that ever since the success of the union, the production of coal had fallen off by at least a quarter and that this had caused the price of coal to rise at market. (The private records of the cartel, interestingly enough, reveal that this certainly was not true. Their own figures state that total production fell by less than 5 percent during the calendar year 1831 as compared with 1830.)[30] The owners claimed further that they had tried to maintain production at 1830 levels in order to keep prices down but that they had been prevented from doing so by the union. The union, the coal owners claimed, had refused to allow extra "strangers" to be employed in the mines, who were needed to increase production. This was the crux of the report: The union had successfully controlled the labor market. The owners warned their audience that they considered this to be illegal and vowed to indemnify any owner who broke the "secret combination to control the free circulation of labour."[31]

Control of the labor market thus became the focal point for the resumption of the conflict between capital and labor in 1832. The coal owners, now eschewing the discourse of master and servant and using exclusively the language of political economy, resolved

> that the experience of the last two years has proved that the Peace and Pros-
> perity of the Coal districts of the North, and the rights and interests of the
> Individual Coalowners, can only be secured by the establishment of a system,

---

[27] Ibid., 17 September 1831, 333.
[28] Ibid., 31 December 1831, 343–4; ibid., 19 January 1832, 349; ibid., 14 February 1832.
[29] Ibid., 21 January 1832, 350–1.
[30] Ibid., 6 January 1832, 346. The production totals fell from 1,184,143 chs. in 1830 to 1,125,892 chs. in 1831.
[31] *Newcastle Chronicle*, 17 March 1832.

under which a constant and regular supply of labour equal to the demand can be promptly procured.[32]

General Bouverie, commander of the Northern District, pointedly explained to Lord Melbourne at the Home Office that control of the market for labor was the critical issue between the miners and their employers:

> It appears that the object of the union of colliers is not so much to enforce any increase of wages, as to prevent the masters from employing any men who from not belonging to the union, as from any other causes, are obnoxious to them, and that the masters are determined to resist this to the utmost.[33]

The binding in April of that year was initially marked by confusion on both sides. About one-half of the collieries in the northern coalfield agreed to renew the bonds of their miners while the remaining owners resisted. This seems to have caused some unease among the miners, and Hepburn organized a mass rally specifically to confirm the fact that "it was agreed among them that such of them as could get their rights should bind; and such of them as could not obtain these rights should remain unbound until they were conceded to them."[34] At several collieries, particularly those that had experienced labor disputes through the winter, lead miners from Weardale were immediately hired to replace union workers.[35] According to the renegotiated terms of the 1831 bond, the miners were allowed a fortnight to leave their homes after the bonds had run out. Consequently, on or about 19 April, owners began forcibly to evict union miners from company-owned houses. The first evictions occurred at Hetton Colliery, and in the succeeding fortnight further evictions were executed at Jarrow, Burdon Main, Townley Main, and Elswick collieries.[36]

Meanwhile, the coal owners took actions designed to more subtly undermine the financial basis of the union. Immediately upon the cessation of the old bonds, the cartel recommended that each colliery limit the earnings of their workers to three shillings a day.[37] It is unclear, however, whether this resolution was at first a conscious effort to restrict the flow of funds to the union or a direct reaction to a new movement of restriction among the union rank and file. Buddle was of the opinion that the union had initiated the policy restricting earnings to three shillings a day but allowing men to

---

[32] N.R.O., General Meetings of the Coal Owners of the Rivers Tyne & Wear, 19 May 1832, 129.

[33] P.R.O., H.O. 40/30/142, Maj. Gen. Bouverie to Lord Melbourne, 20 April 1832.

[34] *Newcastle Chronicle*, 21 April 1832.

[35] N.E.I.M.M.E., Bell Collection, vol. 11, 418.

[36] Ibid., 421; ibid., 425.

[37] N.R.O., Committee Meetings of the Coal Owners of the Rivers Tyne & Wear, 7 April 1832, 368.

work eleven days each fortnight.[38] "The Pitmen," he wrote, "at all the Collieries, where they are bound, have limited themselves to ¾th work – that is to say, they have limited themselves to 3/. instead of 4/. per day – out of which they are paying 9d. or ¼th of their earnings towards the unbound men."[39]

Yet as in 1831, some rank-and-file actions were spontaneous and often uncoordinated. At Wallsend Colliery, the cashier there reported that the hewers were working to four shillings a day while at surrounding collieries they were working only to three shillings.[40] At Rainton Colliery, about one-half of the hewers worked to four shillings a day, while the remainder limited themselves to three shillings.[41] At this juncture, the union leaders' dedication to popular sovereignty began to fail them. At a rally of the pitmen on 16 June, Hepburn indicated that no broad consensus could be reached on whether to limit earnings to three or four shillings a day and that a decision on the matter had to be deferred.[42] It subsequently was decided to raise the limit back up to four shillings in part, it appears, to maintain the flow of income to the union.[43]

Finally, on 12 May, the coal owners agreed to hire only those men who would swear to an oath that they did not belong to a union or any other organization that "prevented [the employee] from the strict performance of any Contract."[44] This amounted to a formal acknowledgment of the union's control of the movement of labor and output. While the oath was certainly employed as a pretense to victimize trade unionists, as it was elsewhere,[45] it also reiterated the owners' instrumental definition of market relations. For the ostensibly "free" worker, *les dés sont pipés*.

By the beginning of June, the union had adumbrated their tactics for the coming months.[46] Hepburn claimed that the miners who were employed could continue to afford to support the half who were on strike for a further ten weeks. During that time, the expenses that were being incurred by the coal owners, especially to hire blacklegs and the police or special constables

---

[38] D.C.R.O., Londonderry Papers, D/Lo/C 142(853), Buddle to Londonderry, 24 June 1832.
[39] Ibid., D/Lo/C 142(794), Buddle to Londonderry, 27 April 1832.
[40] D.C.R.O., John Buddle Papers, NCB I/JB/1188, John Reay to Buddle, 2 April 1832.
[41] D.C.R.O., Londonderry Papers, D/Lo/C 142(831), Buddle to Londonderry, 30 May 1832.
[42] *Newcastle Chronicle*, 23 June 1832.
[43] D.C.R.O., Londonderry Papers, D/Lo/C 142(853), Buddle to Londonderry, 24 June 1832.
[44] N.R.O., Committee Meetings of the Coal Owners of the Rivers Tyne & Wear, 12 May 1832, 374.
[45] On the use of "the document" as a form of blacklisting in textiles, see R. G. Kirby and A. E. Musson, *The Voice of the People: John Doherty, 1798–1854, Trade Unionist, Radical and Factory Reformer* (Manchester: Manchester University Press, 1975).
[46] The following account differs in detail from that provided by Colls in *Pitmen of the Northern Coalfield*, 97.

to protect them, would force them eventually to submit to the union. After all, Hepburn said, "the pitmen were poor, and only could be poor; but the owners were rich, and could be made poor." By September, he suggested that the coal trade would be active again in preparation for winter's demand and "if they were not agreed by that time, he thought it would be advisable for them to knock off altogether." However, Hepburn himself seems to have equivocated on the decision to strike. He made the clear point that this was not his opinion but that "it was the spirit of two-thirds of the men. His advice was, that they should at least wait till that time."[47]

The decision to maintain the status quo for a further ten weeks was a crucial one, as Colls has maintained,[48] because it gave the owners the opportunity to coordinate their actions. In response, the coal owners decided to defeat the union by attrition. Hewers were limited to earning three shillings a day, and the number of days' work was limited to five each week. Thus, as Buddle wrote,

> it was resolved to put them [the hewers] down to the lowest scale – 30s. per Fort. At this rate, viz. 15/. a week, they can at 3d. in the shilling only pay ⅜ each week to the Union – leaving the workers 11s/3d a week to live on. But at the rate at which the *Cock Parlt.* have allowed – 11 days a fortnight at 4/. a day, they could earn 44/. a fortnight or 22/. a week, which together with the money made by their wives, and families working for the Farmers, and the assistance they receive from the Trade's Union, would enable them to hold out well enough for the time they calculate.[49]

Further, the coal owners now drew upon sources of both capital and state power to support their efforts. On 29 May, they approved a resolution to raise £10,000 to be used to hire and relocate new workmen, which then was secured from Ridley's Bank in Newcastle on 10 June.[50] Hiring teams scoured the coalfields of Britain offering up to £3 a head for laborers. "Strangers" were lured from as far away as Wales, Cornwall, Derbyshire, Staffordshire, and Somersetshire to work in the Northeast.[51] Of equal significance was the fact that the military was now convinced that their presence was necessary to maintain the peace. Unlike the strike in 1831 when General Bouverie had decided to leave the coal owners to their own resources, the evictions

[47] D.C.R.O., Londonderry Papers, D/Lo/C 142(831), Buddle to Londonderry, 30 May 1832; D/Lo/C 142(847), 17 June 1832; *Newcastle Chronicle*, 23 June 1832.

[48] Colls, *Pitmen of the Northern Coalfield*, 96–7.

[49] D.C.R.O., Londonderry Papers, D/Lo/C 142(853), Buddle to Londonderry, 24 June 1832.

[50] N.R.O., General Meetings of the Coal Owners of the Rivers Tyne & Wear, 29 May 1832, 131; D.C.R.O., Londonderry Papers, D/Lo/C 142(840), Buddle to Londonderry, 10 June 1832.

[51] D.C.R.O., Londonderry Papers, D/Lo/C 142(836), Buddle to Londonderry, 4 June 1832; D/Lo/C 142(855), 27 June 1832; D/Lo/C 142(858), 29 June 1832.

of 1832 were accompanied by a substantial military presence. Cavalry and infantry were stationed at Chester-le-Street, Houghton-le-Spring, South Shields, Jarrow, Hetton-le-Hole, North Shields, and Newcastle.[52] Bouverie encouraged the local magistrates to form a permanent police force and agreed to arm special constables with about one hundred carbines and muskets.[53] By mid-June, he realized that the union would be defeated only by "the starvation of the Pitmen" and that the coal owners' lives and property were secured only by "the almost military occupation of the Collieries."[54]

Interestingly, the despised Lord Londonderry did not subscribe either to the bank loan or to the earnings' limit on hewers. Buddle, in fact, was forced to sneak out of the meeting at which the men's work was reduced to fifteen shillings a week because he knew Londonderry would not consent to that limitation.[55] Buddle also reported a rumor that the Londonderry collieries would be exempted from a new strike by the union "because he is not in the Coal-owners' Union."[56] (In the midst of the 1844 miners' strike, William Mitchell, a trade union pamphleteer, fondly recalled 1831–2, when "the Marquis of Londonderry, judging for himself, and resolving not to be hood-winked by the others, broke from your union [i.e., cartel], conceded the principal portion of our claims, and thus compelled the remainder of your body to come to a similar agreement.")[57] Londonderry's finances were in particularly poor shape at this time, especially due to the heavy outlays on Seaham Harbor, and he could not risk labor problems at his works. At the coal owners' meeting, Buddle tried to make it clear that he desperately wanted to fight the union, but "of course, I could not hint at the *true cause* why yr. Lordship does not concur in the anti-union plans."[58]

Refusing to sanction violence or any form of picketing, "flying" or otherwise, the union sent two delegates to the south of England to publicize their position and discourage other "strangers."[59] Predictably, this had little effect, and the angry mood of some of the rank and file became increasingly apparent. In May, shots were fired into the homes of nonunion miners at Mount Moor and Black Fell collieries.[60] In June, two miners were sentenced

[52] P.R.O., H.O. 40/30/144, 2 May 1832.
[53] Ibid.
[54] P.R.O., H.O., 40/30/176, Bouverie to Maj. Gen. Lord Fitzroy Somerset, 17 June 1832.
[55] D.C.R.O., Londonderry Papers, D/Lo/C 142(831), Buddle to Londonderry, 30 May 1832;
D/Lo/C 142(855), Buddle to Londonderry, 27 June 1832.
[56] Ibid., D/Lo/C 142(820), Buddle to Londonderry, 23 May 1832.
[57] Mitchell, *The Question Answered*, 21.
[58] D.C.R.O., Londonderry Papers, D/Lo/C 142(857), Buddle to Londonderry, 28 June 1832.
[59] *Newcastle Chronicle*, 23 June 1832; D.C.R.O., Londonderry Papers, D/Lo/C 142(851),
Buddle to Londonderry, 21 June 1832.
[60] *Newcastle Courant*, 26 May 1832; ibid., 28 April 1832.

to six months' hard labor for threatening blacklegs at Burdon Main Colliery.[61] In July, two special constables were attacked after they tried to prevent union miners from harassing blacklegs at Hetton Colliery.[62] Finally, the approbation of the public that was such an important aspect of union tactics suffered irrevocably when Nicholas Fairless, an eighty-year-old magistrate from South Shields, was murdered by two striking miners.[63] Needless to say, criminal justice was as difficult for the union to secure as economic justice. A constable was convicted only of manslaughter in the murder of a union pitman at Chirton Colliery and was sentenced to six months in prison.[64] Fairless's murderer, William Jobling, however, "was hung, covered in pitch, encased in iron stirrups and bars, and then re-hung in chains at the scene of the crime."[65]

Hepburn nevertheless remained dedicated to the policies of moderation and negotiation. In late June, he told a pitmen's rally that "*he* and a deputation of delegates, meant to go to the Coal-trade office ... to demand a meeting with a deputation of Viewers and Coal-owners – to talk matters over, and to shew the latter how very far *he* could beat them, at all points – in reasoning and argument."[66] Yet caught in the vise of restricted production, fewer employed union members, and rising expenditures to support the unemployed, union funds dwindled. In July, the delegates were forced to raise weekly contributions from four to six shillings a man. More and more miners were forced to beg and steal to survive. Two pitmen were convicted of begging in Newcastle and sentenced to a month's hard labor.[67] Union miners who had been evicted from their homes camped out along the roadsides and, according to Henry Morton, filled Methodist chapels.[68] By mid-July, General Bouverie estimated that there were ten thousand unemployed and homeless miners and their families.[69]

At the beginning of August, the delegates recognized that their position was so weakened that only an immediate strike of all collieries could hope to save the union. The delegates to the union were asked to poll their members. If the delegates failed to get support for a strike, they had decided to keep on for a month longer and try to gain a negotiated settlement.[70] The

---

[61] *Newcastle Chronicle,* 9 June 1832.
[62] N.E.I.M.M.E., Bell Collection, vol. 11, 539.
[63] *Newcastle Chronicle,* 16 June 1832; ibid., 30 June 1832.
[64] D.C.R.O., Londonderry Papers, D/Lo/C 142(864), Buddle to Londonderry, 4 August 1832.
[65] Colls, *Pitmen of the Northern Coalfield,* 252.
[66] D.C.R.O., Londonderry Papers, D/Lo/C 142(855), Buddle to Londonderry, 27 June 1832.
[67] *Newcastle Courant,* 9 June 1832.
[68] Lambton Mss., Morton to Durham, 22 May 1832.
[69] P.R.O., H.O., 40/30/190, Bouverie to Melbourne, 16 July 1832.
[70] D.C.R.O., Londonderry Papers, D/Lo/C 142(866), Buddle to Londonderry, 9 August 1832.

rank and file undoubtedly voted against a strike; for on 10 August, Robert Lowery, the North Shields radical, was enlisted to draft a letter to R. W. Brandling, the chairman of the coal owners' cartel, requesting a conference to resolve the dispute.[71] As Buddle noted, from the receipt of the letter "we can infer that the Delegates have not been able to carry their plan of the General Stop."[72] On 21 August, the committee of coal owners resolved that "no beneficial result can ... be expected from any negociation [*sic*] with the Delegates or Managers of the Pitmens Union."[73]

The union then began to break apart. Over a hundred men at Lord Durham's collieries refused to contribute any more to the movement.[74] And when the union supporters at Tommy Hepburn's Hetton Colliery sued for work, this signaled the end of the movement. Miners throughout the coalfield subsequently followed suit.[75] Hepburn himself continued to lead the rump of the union into early October, but the union had been defeated.

The experience of Hepburn's union movement left a double legacy to the northern miners. To some, the brief success of unionization proved that grievances could be successfully redressed within the existing structure of economic and political relations. To others, however, the failure of the union illustrated the inseparability of economic justice from political equality. The two legacies were certainly neither antithetical nor mutually exclusive. Some people, notably Benjamin Embleton, were active in both trade unionism and Chartism for many years.[76] However, others such as Tommy Hepburn abjured trade unionism in preference to an active role in the Northern Political Union. Benjamin Pyle, in contrast, appears to have followed the opposite course and remained dedicated to the efficacy of economism.[77]

Hepburn's role in the Northeast after the defeat of the union changed dramatically and exemplified the political course of action. By 1838, he was active in the anti-Poor Law movement and was an early Chartist.[78] He eschewed his former discourse steeped in Hodgkins and Thompson and later employed the radical political language of "natural rights," something that had not occurred during his strike movement. In June 1838, for example,

---

[71] Ibid., D/Lo/C 142(866), Buddle to Londonderry, 11 August 1832.

[72] Ibid.

[73] N.R.O., Committee Meetings of the Coal Owners of the Rivers Tyne & Wear, 21 August 1832, 399.

[74] Hammond and Hammond, *The Skilled Labourer*, 35.

[75] N.E.I.M.M.E., Bell Collection, vol. 11, 543.

[76] Challinor and Ripley rightly claim that Embleton is "one of the unsung heroes of the trade union movement" in *Miners' Association*, 249.

[77] At a union rally in Northumberland in 1843, Pyle was accused of claiming that the Chartists were destructive; see Challinor and Ripley, *Miners' Association*, 19.

[78] *Northern Star, and Leeds General Advertiser*, 1 January 1838.

he moved a resolution at a universal suffrage rally in Newcastle to the effect that "the only security against the corruption of the few, and the degradation of the many, is to give the people equal rights, social and political." To obtain these rights, Hepburn now advocated universal suffrage, secret ballot, annual elections, and the abolition of all qualifications to be a Member of Parliament.[79] In May 1839, he chaired a "monster" Chartist rally in Newcastle where he drew on some of his old vocabulary of mutuality, order, and determination, but now saw himself as a political radical and claimed that "all that radicals wish is their natural right, and when that is obtained, their peaceful conduct will shame their oppressors."[80]

While the language and ideology of trade unionism after 1832 similarly reflected a more vigorous conception of natural rights,[81] it also continued to bear the profound imprint of Ricardian socialism, especially those aspects that elaborated a conception of the perfection of market mechanisms. The *Miners' Advocate* of the 1840s, for example, continued to argue for the reciprocal power of capital and labor: "We are not so ignorant as to believe that he who sinks his capital should not be remunerated.... But while we admit this, we claim the same protection for our capital, namely our labour."[82] Similarly, the succeeding generation of miners' leaders accepted the perfectibility of market relations under capitalism that would benefit both capital and labor: "We unite to reduce the hours of labour, to call into employ the unemployed, and ultimately to make the product of our labour scarce in the market, and thus give an advance of profit to the employer and better wages to the workmen."[83] It was along this road that the sliding scale and district-wide conciliation boards of the later nineteenth century clearly lay.

Mid-Victorian reformism and the objectification of market relations

The achievements of mid-Victorian miners' unionism clearly continued to elaborate principles similar, if not identical, to those of trade unionists prior to 1850. Welbourne's claim, reflecting the Webbs' contention, that the decades surrounding midcentury marked a transition from customary industrial relations based upon the laws governing masters and servants to that of modern

[79] Ibid., 30 June 1838.
[80] Ibid., 25 May 1839.
[81] Colls, *Pitmen of the Northern Coalfield*, 296. However, Colls's claim that this " 'labourist' political economy [was] posed mainly against the market" cannot be supported; see ibid., 291.
[82] Quoted in ibid., 290.
[83] Ibid.

collective bargaining clearly cannot be supported.[84] Collective bargaining may have developed different characteristics, but it was not a significant departure from earlier forms of industrial relations. Similarly, Keith Burgess's thesis that the post-1850 policies of accommodation were the legacy of the prosperous conditions of the third quarter of the nineteenth century that allowed for the bureaucratization of trade unionism and the consequent incorporation of reformist union officials merits review and substantive revision.[85] Both the theoretical assumptions as well as the evidentiary basis of Burgess's claims have recently been seriously called into question.[86]

Instead, there was a marked continuity in the industrial relations of the northern coalfield between 1800 and 1875 that was characterized by attempts to control the forces of the market for labor and products. Throughout this era, bargaining was accepted as the principal terrain of labor relations. Moreover, forms of arbitration and conciliation were continually employed and adapted. While it would not be strictly accurate to claim that arbitration and conciliation became formalized after midcentury – as we have seen, they were institutionalized in the annual bond as early as 1810 – it is true, nonetheless, that before 1850 practice never fully accorded with theory. Arbitration was employed sporadically and without uniformity. Most negotiations occurred without third parties, directly between the viewers and the men, both of whom largely accepted the ethic and culture of bargaining.[87]

During the half-century after 1825, the focus of trade unionism became trained ever increasingly upon the inequitable powers retained by the viewer, or later the colliery manager. Several of the most noted achievements of trade unionism in the northern coalfield during the third quarter of the nineteenth century, such as the institution of checkweighmen, the acceptance of formal arbitration, and the adoption of the sliding scale, can be largely construed as part of a continuous effort to restrict managerial authority and, thereby, to augment the market power of labor. The most notable attempts

---

[84] Welbourne, *Miners' Unions of Northumberland and Durham*, 26–7, cited in Church, *History of the British Coal Industry*, 677.

[85] Keith Burgess, *The Origins of British Industrial Relations* (London: Croom Helm, 1975), 172, 175–6, 184–91.

[86] The fundamental prosperity of these years has been questioned by Church, who also claims that there is little evidence of a marked division between union leaders and the rank and file at least before 1880, and in Durham before the 1892 strike; see Church, *History of the British Coal Industry*, 702–13. For a general critique of Burgess's "rank-and-filism," see Jonathan Zeitlin, "From Labour History to the History of Industrial Relations," *Economic History Review*, 2nd ser., 40, no. 2 (1987), 159–84.

[87] Thus, the distinction between substantive and procedural rule-making systems employed by Price does not adequately reflect the amalgam that constituted industrial relations in the northern coal industry; see Price, *Masters, Unions and Men*, 73–4.

in this area were the efforts to rationalize both market forces and institutional authority by creating objective determinants of wages and earnings. It was these attempts to objectify market relations at the pithead that marked the contribution of trade unionism in the late nineteenth century.

The 1860 Mines Regulation Act was perhaps the first such attempt. It sought to do away with measurement and payment of hewers by volume and to substitute in its stead payment by weight. In addition, the act provided for the appointment of checkweighmen from among the employees of the colliery to supervise the weighing of coal.[88] The substitution of payment by weight instead of measure had been advocated by trade unionists from as early as 1847 when Martin Jude's petition to the House of Commons demanded parliamentary legislation on this matter.[89] Certainly before that, however, the laying out of entire corves and the application of fines for underfilled and foul (mixed with dirt, stone, or small coal) corves had been among the most prominent complaints of the union movements of the first half of the century.[90]

It is somewhat surprising that the role of the checkweighman has garnered more significant attention in the history of industrial relations than the signal importance of payment by weight. The Webbs were the first to argue that the position of checkweighman became "an admirable recruiting ground from which a practically inexhaustible supply of efficient Trade Union secretaries or labour representatives can be drawn."[91] This contention was later contested by Keith Burgess, however, who suggested that the position drove a wedge among the rank and file and created a level of trade union officialdom that helped to blunt the impact of popular activism. Checkweighmen, he claimed, adopted reformism largely because their earnings were more secure than hewers and other workers, their position was in part dependent upon the acquiescence and acceptance of employers (at least until 1887 when the checkweighmen's independence was finally secured), and their privileged position might be threatened by strikes or work stoppages.[92]

The social position of the checkweighman is an interesting question, and the difference in interpretation may be explained in part by the periodization of the problem. John Wilson, the Durham miners' leader, believed that before 1887, the checkweighman generally was the employer's man because his

---

[88] Webb and Webb, *History of Trade Unionism*, 289–91; Church, *History of the British Coal Industry*, 702–4; Burgess, *Origins of British Industrial Relations*, 166–7.

[89] Fynes, *Miners of Northumberland and Durham*, 119–21.

[90] On the 1825 United Colliers, see *A Voice from the Coal Mines*; on Hepburn's union, see above, Chapter 7; for the Miners' Association of the 1840s, see Challinor and Ripley, *Miners' Association*, 99–110.

[91] Webb and Webb, *History of Trade Unionism*, 291.

[92] Burgess, *Origins of British Industrial Relations*, 167.

appointment was subject to the approval of the owner. After the passage of revisions in the Coal Mines Regulation Act, the checkweighman became independent of managerial authority. After 1887, checkweighmen were "employed by the workmen" and became, in Wilson's words, "the mouth-piece of the men when meeting the manager [and] the leader in public movements."[93]

Ultimately, however, the role of the checkweighman may have been less significant to the history of industrial relations in the northern coalfield than the transition to payment by weight after 1860. With the introduction of payment by weight, the viewer or manager's autonomous control (the colliers repeatedly referring to it as "despotic" or "tyrannical" control) of the measurement and quality of production was compromised and undermined. Admittedly, to some extent, grievances concerning faulty weighing machines and disputes over payment by weight replaced the older complaints concerning fines and the size of corves.[94] Nonetheless, managerial authority was significantly circumscribed by the 1860, 1872, and 1887 acts; the manager's ability to victimize, intimidate, and exploit miners through the abuse of fining and exacting overmeasure was limited. This obviated the cause of many of the most famous colliery strikes of the mid-nineteenth century, like that at Thornley Colliery in 1844[95] and the "Rocking Tub" strike in 1863,[96] that had been provoked by the patent exploitation of hewers through the system of fining and measurement by volume. In a sense, the checkweighmen were merely symbols of a new balance of power. After the implementation of payment by weight, there appeared to be a more significant reciprocity of power between capital and labor at the pithead, especially since the measurement of production was delegated to the objective balance of the weighing machine.

In an even more pronounced manner, the introduction of formal arbitration and sliding wage scales tied to prices combined traditional attitudes toward arbitration, the struggle for new forms of market power, and an attempt to objectify social relations at the pithead.[97] The 1871 decision of the Durham Miners' Association under William Crawford to propose the establishment of a board of arbitration to settle a dispute at Thornley Colliery and the subsequent formation of the Joint Committee of coal owners and

---

[93] John Wilson, *A History of the Durham Miners' Association, 1870–1904* (Durham Veitch, 1907), 338–9.
[94] Church, *History of the British Coal Industry*, 703.
[95] Challinor and Ripley, *Miners' Association*, 99–106.
[96] Welbourne, *Miners' Unions of Northumberland and Durham*, 115–21; Fynes, *Miners of Northumberland and Durham*, 226–9.
[97] J. H. Porter, "Wage Bargaining under Conciliation Agreements, 1860–1914," *Economic History Review*, 2nd ser., 23, no. 3 (1970), 460–75; Church, *History*, 607–701.

trade unionists to regulate wages and arbitrate disputes throughout the
county in the following year bears a remarkable resemblance to earlier forms
of arbitration.[98] While the composition of the Joint Committee differed
significantly from earlier forums in that the union now possessed representation on the arbitration panel (whereas previously only viewers had arbitrated
wage bargains), the process of arbitration clearly echoed past experience.
Notably, if the Joint Committee was unable to reach a decision, it had the
authority to refer matters to two arbitrators who then appointed an umpire.
A similar format of arbitration had already been included in the bonds after
1810 and advocated by the United Colliers in 1825.[99]

The willingness of the coal owners to participate in a formal system of
arbitration may have been linked to the realization of their declining power
in the market at precisely the time the industry experienced a "colossal price
inflation."[100] After 1872, however, arbitration came under increasing attack
for its failure to adequately defend the interests of the rank and file,[101] and
it was replaced several years later by the sliding scale introduced among the
Northumberland collieries in 1877. The sliding scale, which tied wages to
the price of coal, was an explicit acceptance of the commodity market's
determination of wages as well as the continued acknowledgment of, in the
words of the Durham miners' leader William Crawford, "a respect for the
rights . . . of capital."[102] However, like payment by weight, the sliding scale
sought to establish a greater degree of equity in exchange relations at the
colliery level by the objectification of market relations. Managerial authority
over on-site bargaining of piece rates was circumscribed, and the wage
bargain was thereafter supposed to be determined largely through regular
district-wide negotiations over base rates.[103] Bargaining at the point of production, Rowe was careful to point out, did not disappear, but it came to
operate as a subtext to regional wage settlements.[104]

The policies of union leaders, such as William Crawford, John Wilson
and Thomas Burt, therefore, cannot be labeled simply as accommodationist
or incorporationist, implying, as it does, that they reflected a unique stage in
the development of the working class under industrial capitalism. Instead,
those policies reflected the deep-seated acceptance of market relations, wage
bargaining, and forms of arbitration and conciliation that can be traced back

[98] Wilson, *A History of the Durham Miners' Association*, 32–4, 66–71.
[99] See Chapter 5.
[100] Church, *History of the British Coal Industry*, 55.
[101] Welbourne, *Miners' Unions of Northumberland and Durham*, 185–6; Porter, "Wage Bargaining," 465.
[102] Welbourne, *Miners' Unions of Northumberland and Durham*, 157.
[103] Porter, "Wage Bargaining," 467–8.
[104] Rowe, *Wages in the Coal Industry*, 49–50.

at least to the beginning of the nineteenth century. More significantly, how-ever, the accommodationist tactics of the late-nineteenth-century miners' leaders may more accurately be described as an attempt to objectify market relations and thus to create a system of more equitable exchanges. In their attempt to articulate these assumptions, these trade union leaders drew on a tradition that believed, in Wilson's words, that "reciprocity and mutuality form the platform upon which the two sides [i.e., capital and labor] can meet."[105] These assumptions, rooted as they were in both the labor process and the system of industrial relations of the early nineteenth century, were variations of an earlier generation's struggle for market power.

[105] Wilson, *A History of the Durham Miners' Association,* 339.

Conclusion: the labor process and the market

Without exchanges there can be no industry, no continued production of wealth. Labor without exchanges would be nearly as useless as exchanges without labor.

> William Thompson, *An Inquiry into the Principles*
> *of the Distribution of Wealth Most Conducive*
> *to Human Happiness* (1824)

Social historians writing about the early nineteenth century have tended to erect the acceptance of the market as part of a litmus test of cultural identity and class relations. It has been argued, for example, that the common theme running through E. P. Thompson's histories has been that of class struggle as expressed through antithetical cultures.[1] Thus, Thompson has claimed that the working and middle classes were distinguished by their "alternative and irreconcilable views of the human order – one based on mutuality, the other on competition."[2] The moral economy of the English crowd, of course, was the antithesis of the political economy of the bourgeoisie.[3] In a similar manner, E. J. Hobsbawm, although more circumspect of the notion of class-specific acceptance of market relations, nevertheless charted the mid-nineteenth century as the era during which skilled workers eschewed their customary, nonmarket attitudes toward work and learned "the rules of the game."[4] Among French social historians, William Reddy has been most prominent in applying a cultural paradigm of class struggle to the French

---

[1] Richard Johnson, "Three Problematics: Elements of a Theory of Working-class Culture," in Clarke, Critcher, and Johnson, *Working Class Culture*, 220–1.

[2] Thompson, *Making of the English Working Class*, 206.

[3] Idem, "The Moral Economy of the English Crowd in the Eighteenth Century," *Past & Present*, no. 50 (February 1971), 76–136.

[4] E. J. Hobsbawm, "Custom, Wages and Work-load in Nineteenth-century Industry," *Essays in Labour History*, Asa Briggs and John Saville (eds.) (London: Macmillan, 1960), 113–39. On the dichotomy between custom and market relations, see also Richard Price, "Structures of Subordination in Nineteenth-Century British Industry," in Pat Thane, Geoffrey Crossick, and Roderick Floud (eds.), *The Power of the Past* (Cambridge University Press, 1984), 122–3.

196

textile trades, claiming to see "the whole range of political upheavals of the first half of the nineteenth century as a confrontation between two alternative cultural orders."[5]

The dichotomy between "custom" and "market," like the antithesis of individualism and mutuality, is to a certain extent the legacy of nineteenth-century sociology.[6] Its purported value is to reveal the qualitative difference between the preindustrial artisanal community and the competitive individualism of modern capitalist society. Moreover, it attempts to resolve the perceived dichotomy in British history whereby the working class has developed a culture that is both independent, or hermetic, and subordinate.[7] However, I have tried to argue that the analytical boundaries separating customary from market relations are not applicable without significant revision. Moreover, the distinction between custom and market has served to obscure not only the nature of class relations under early industrial capitalism, but also the character of the labor process and the sources of class conflict.

Market relations penetrated local society principally through experience gained in the workplace. There, bargaining over piece rates occurred regularly between management and adult male workers and was accepted as the principal terrain of industrial relations. In this sense, P. K. Edwards is certainly correct to argue that "the capitalist labor process is necessarily closely connected with the capitalist market."[8] However, contra Edwards, the labor process is not tied to the market solely at the macroeconomic level of employment, the location of industry, or the availability of commodity markets in which surplus value may be realized. Instead, as D. J. Lee has pointed out, if one accepts Baldamus's claim that the labor process is characterized by the indeterminancy of the labor contract – that is, the uncertainty over how the employer's instructions and rules will be implemented or amended by the worker at the point of production – then one must realize as well that the labor process itself is a market process. Baldamus's "effort

---

[5] Reddy, *Rise of Market Culture*, 15.

[6] On both the theoretical and empirical weaknesses of the antithesis between custom and market in the case of the French trades of the eighteenth century, see Michael Sonenscher, *Hatters of Eighteenth Century France*, 18–19; idem, *Work and Wages: Natural Law, Politics and the Eighteenth-century French Trades* (Cambridge University Press, 1989), 44–6 and chap. 6. See also Gerald M. Sider, "The Ties That Bind: Culture and Agriculture, Property and Propriety in the Newfoundland Village Fishery," *Social History*, 5, no. 1 (1980), 1–39.

[7] J. M. Cousins and R. L. Davis, "'Working Class Incorporation' – A Historical Approach with Reference to the Mining Communities of S. E. Northumberland, 1840–1890," in Frank Parkin (ed.), *The Social Analysis of Class Structure* (London: Tavistock, 1974), 275–8. However, Cousins and Davis's argument that working-class reformism was based upon the "mutual dependency" of capital and labor in the face of hostile markets is not supported here.

[8] Edwards, *Conflict at Work*, 67.

bargain" entails the constant evaluation and reevaluation of the labor contract at the point of production on the basis of market power.[9]

Within the northern coal industry, the ubiquity of bargaining was perhaps the most important feature that negotiated relations between classes. Moreover, it was the foundation upon which were built the sliding scale and district-wide settlements of the later nineteenth century. Thus, the origins of collective bargaining lay not in the collusion of employers with accommodative trade union officials after 1850,[10] but in the history and culture of work.

Significantly, working for piece rates and acquiring the ability to bargain, and bargain well, took on broad cultural ramifications. Piece rates not only signified the workers' autonomy and independence, as Marx noted, but in the coal industry, bargaining for piece rates was a symbol of passage into adulthood and a test of masculinity. Paradoxically, as with many male contests, bargaining could also be construed as an element of camaraderie that cut across class lines: Tough bargaining was celebrated as a "good fight." In this sense, the culture of work in the mines reflected the perception of the labor process as a game. And as Michael Burawoy has written, it is the widespread acceptance of this shopfloor game that can serve to generate consent to the general rules of capitalist production.[11]

While the game metaphor should not be forced to bear too much of the burden of explaining the accommodation of the working class to industrial capitalism, it can provide some insight into the structuring of class relations. The game metaphor assumes that while the worker objectively is subordinated to the production process, the acceptance of the game's rules at the point of production obscures this fact to the workers. Thus, the labor-process game masks the inequities of capitalism by a system that amounts to little more than diversion.[12]

Why would a worker participate in such a game? Coercion is certainly one element. However, coercion alone is inefficient and, it has been argued,

---

[9] D. J. Lee, "Skill, Craft and Class: A Theoretical Critique and a Critical Case," *Sociology*, 15, no. 1 (1981), 60; on the significance of Baldamus's work for Edwards, see *Conflict at Work*, 32–5; Paul Thompson, *The Nature of Work: An Introduction to Debates on the Labour Process* (London: Macmillan, 1983), 101–6.

[10] For a further example of this line of thought, see Clegg, Fox, and Thompson, *History of British Trade Unions*, 1: 23–4.

[11] Michael Burawoy, *Manufacturing Consent: Changes in the Labour Process under Monopoly Capitalism* (Chicago: University of Chicago Press, 1979), 77–94. The nature of this game, however, is markedly different than that described by Burawoy. He elaborated a theory in which the labor process game took place within the context of externally imposed rules of production. Conversely, the coal miners' game could be thought of as concerning the formulation of the rules themselves.

[12] Ibid., 77–82.

cannot secure the active participation and consent of workers that is necessary for capitalism's success.[13] It has been suggested that the acceptance of the rules of the game are brought about by the social pressure on the shopfloor to "fit in" and, more importantly, by the desire to "make out," that is, to manipulate the rules of work in order to complete a task or quota as quickly as possible.[14] Yet this apparently begs the question: If the work rules are externally imposed without consent, why are the rules themselves, as opposed to the shop culture in which they are imbedded, so readily accepted other than because workers fear the coercive power of capital?

Part of the answer may lie in the fact that these rules of the game themselves appear to be, or can be made to appear to be, fair and equitable despite the worker's surbordinate role in the production process. Thus, in the nineteenth-century coal industry, piece work appeared to link both personal income and profits to commodity markets, an apparently objective determinant of wages as long as the commodity markets were not "unfairly" manipulated by the collusion of coal owners. Similarly, bargaining could be accepted as a form of industrial relations as long as labor and capital achieved the appearance of parity. Contrary, therefore, to the game metaphor, the acceptance of these rules does not entail the acceptance of capitalism *tout court*.[15] Instead, as in the case of the northern coal miners, the acceptance of these rules indicated the internalization of a set of ideals that to a certain extent were compatible with, but not necessarily identical to, capitalist ideology. These ideals were based on the acceptance of market principles provided that market relations expressed equal and reciprocal power over capital and labor. Thus, in the early-nineteenth-century northern coal industry, the acceptance of piece work and wage bargaining did not entail the full acceptance of the distribution of political and economic power under early industrial capitalism. Bargaining expressed the vision of a social order built on equitable exchange relations. Similarly, unionization was premised upon the construction of a new set of market relations that was deemed fair and equitable. A market culture, therefore, was not a set of wrong perceptions or misguided practices;[16] on the contrary, market culture was a crucial component of working-class culture.

Perhaps the most significant methodological problem raised by the game-

---

[13] Ibid., 83; Edwards, *Conflict at Work*, 49–50.

[14] Burawoy, *Manufacturing Consent*, 51, 64.

[15] Edwards, *Conflict at Work*, 50; Michael Mann, "The Social Cohesion of Liberal Democracy," *American Sociological Review*, 35, no. 3 (1970), 423–39. Patrick Joyce has made a related point in "The Historical Meanings of Work: An Introduction" that workers may have opposed certain aspects of the capitalist system without rejecting the market system *per se*. See Joyce, *Historical Meanings of Work*, 26–7.

[16] Reddy, *The Rise of Market Culture*, 1–2.

playing analogy is the element of reductionism inherent in its application to capitalist society as a whole.[17] Thus, Burawoy contended that the game metaphor "represents the link between individual rationality and the rationality of the capitalist system."[18] While shopfloor relations certainly reflect social relations under industrial capitalism, and to some extent may even epitomize them, the shopfloor is not the sole terrain on which either social relations or ideology are constituted.[19] This may be the case particularly for industries, such as coal mining in the nineteenth century, in which the relative autonomy of the worker at the point of production was accepted by both management and ownership. In such cases, the frontier of control is not situated on the shopfloor. Instead, social relations and ideology may be constituted through struggles over several territories including, in the case of the northern miners, some aspects of the labor process at the point of production, the disposition of the commodity market, and the organization of the labor market.

That the center of social conflict may be located outside the labor process is, of course, contrary to Burawoy's analysis of the formation of consent as well as to Braverman's influential and important work on labor and the workplace in the development of capitalism.[20] A detailed recapitulation of Braverman's thesis need not detain us here, but suffice it to say that he locates the motive force of capitalism in its drive to secure the "real" subordination of labor through the separation of conception from execution in the labor process, that is, deskilling, and the organization and achievement of effective managerial control of the workplace. Among British labor historians influenced by this thesis, Gareth Stedman Jones has applied Braverman's developmental dichotomy between "formal" and "real" control over the workplace to Britain's Industrial Revolution, while Richard Price, emphasizing the dialectic of conflict over these issues, has argued that the struggle for control over the workplace constitutes the central dynamic of modern British industrial relations.[21]

The reaction to this body of work, however, has produced two related

---

[17] Edwards, *Conflict at Work*, 46–8; Thompson, *The Nature of Work*; see also Littler and Salaman, "Bravermania and Beyond," 251–69.

[18] Burawoy, *Manufacturing Consent*, 92.

[19] Michael Mann, for instance, has argued that there is a fundamental disjunction between working-class concrete experience and abstract values. See Mann, "The Social Cohesion of Liberal Democracy," 429; Edwards, *Conflict at Work*, 50.

[20] Burawoy, *Manufacturing Consent*; Harry Braverman, *Labor and Monopoly Capital* (New York: Monthly Review Press, 1974).

[21] Gareth Stedman Jones, "Class Struggle and the Industrial Revolution," *New Left Review*, vol. 90 (March–April 1975); Price, "The Labour Process and Labour History"; idem, *Masters, Unions and Men*.

themes that are of particular significance to the argument presented here: First, that direct control of the workplace may not be of paramount importance to either management or ownership; and, second, perhaps more importantly, that exploitation need not take place at the point of production.

In the first instance, Burawoy, Friedman, and Littler and Salaman, among others, have shown both that under certain conditions, management may yield authority willingly at the shopfloor level and that managerial strategy often includes a significant degree of accommodation and cooperation to ensure production.[22] In the case of the northern coal industry, both of these insights are more appropriate to the analysis of the historical development of social relations than is the quest for direct control or the dialectic between autonomy and subordination. In the first half of the nineteenth century, managerial practice rejected the feasibility of workplace control and instead relied upon piece rates and an intricate system of fines to ensure production and quality. Furthermore, given the absence of management from the shopfloor, worker participation in nominally managerial decisions was sometimes elicited, and there is evidence that changes in work rules were negotiated.

Correspondingly, as Littler and Salaman hypothesized, it appears that forms of control away from the point of production took on greater significance for management and ownership under these conditions. For example, control of housing and other forms of industrial paternalism were considered essential to the structure of authority as well as the profitability of industry well before the advent of the "New Model" employers.[23] Therefore, the relative absence of conflict over the labor process had already shifted the struggle for control away from the point of production well before midcentury.

In the second instance, that of the locus of exploitation, the past several years have witnessed an interesting reorientation of Marxist inquiry into the materialist basis of class relations. From very different perspectives, historians, economists, and sociologists have helped to reassess the significance of the role of the labor process under capitalism. On the one hand, Noel

---

[22] Burawoy, *Manufacturing Consent*, 60–2; A. Friedman, "Responsible Autonomy versus Direct Control over the Labour Process," *Capital & Class*, no. 1 (Spring 1977), 43–57; Littler and Salaman, "Bravermania and Beyond," 251–69. More recently, P. K. Edwards has argued that the dichotomy erected between resistance and accommodation as keys to understanding the labor process is a spurious one. Instead, Edwards sees the workers' experience in this regard as situational. That is, workers have no inherent interest in resisting control or accommodating it. Instead, they seek to "find means of living with the system as they find it." In trying to do this, workers' actions may be interpreted as either forms of resistance or accommodation (or both, as in the case of the use of "angles" or Burawoy's analysis of "making out"), "but there is no overarching interest" that workers bring to the labor process: *Conflict at Work*, 42–6.

[23] On "New Model" employers, see Joyce, *Work, Society and Politics*, 134–157.

Thompson's and Gregory Claeys's research into nineteenth-century working-class ideology has revealed the preponderance of contemporary interest in elaborating theories of exploitation through inequitable exchange relations rather than Marx's later analysis of exploitation as the extraction of surplus value at the point of production.[24] Despite Thompson's obvious concern to elaborate the "failure" of popular economy in comparison to classical Marxian theory, the popular political economists' interest in articulating exploitation through exchange bears a striking resemblance to the current work of John Roemer and Erik Olin Wright.

Roemer's dense work on exploitation and class has sought to prove that class relationships are determined by the social distribution of property rights and, correspondingly, that exploitation occurs outside of production relationships.[25] As Wright has noted, the importance of Roemer's work is to identify the source of conflict between capital and labor as the unequal distribution of the effective control of society's productive assets. As such, exploitation, that is the extraction and appropriation of "surplus value," occurs through the "mediating mechanism" of market exchanges rather than through capitalist domination at the point of production.[26]

Perhaps not surprisingly, my reservations in regard to Roemer's and Wright's work concerns its validity in the face of empirical historical observation.[27] In this regard, an apparent weakness of their work is its failure to adequately assess the historical contingency of definitions and control of the market. The market does not stand between classes solely as a mechanism of mediation but, as this work has tried to show, is itself the subject of struggles for dominance and power. The precise nature of that struggle is determined by a variety of factors among capital, labor, and the state as well as by forces operating principally among capitalists, among workers, or among state officials.

During the first half of the nineteenth century, market relations were an accepted and fundamental premise of social relations throughout the northern coal industry. The struggles between capital and labor were not those between market rationality and custom; nor were they a process of learning and accepting the "rules of the game"; nor were they fundamentally a struggle over the control of the labor process at the point of production. The struggle between capital and labor concerned the interpretation of the

---

[24] Thompson, *The People's Science*. On Owenite theories of exchange in particular, see Claeys, *Machinery, Money and the Millenium*, and idem, *Citizens and Saints*, 148–50.

[25] John E. Roemer, *A General Theory of Exploitation and Class* (Cambridge, Mass.: Harvard University Press, 1982).

[26] Erik Olin Wright, *Classes* (London: Verso, 1985), 64–73, 82–6.

[27] See also the comments of Edwards, *Conflict at Work*, 11–12, who retains an emphasis upon exploitation through the labor process as opposed to distribution.

meaning and control of the reality of market relations. Both capital and labor expressed an acceptance of market relations that was essentially instrumental; market relations were deemed "fair" when they were suitably structured in one's own best interests. For workers, this entailed defense of wage bargaining, union control of the labor market, and a competitive commodity market. For capital, the construction of favorable market relations meant a "free" labor market, oligopolistic control of product markets, and limited wage bargaining. The advent of industrial capitalism in Britain, therefore, did not usher in an era of conflict between customary and market ideologies or the inexorable movement from formal to real control of the workplace; instead, both real and ideological control of the market was the focus of conflict between capital and labor.

Appendix

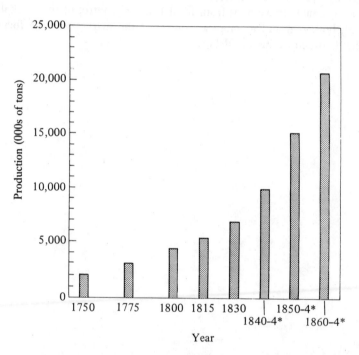

Figure A.1. Estimated Northeast coal production, 1750–1865. Asterisk indicates average per annum. (From Michael W. Flinn, *The History of the British Coal Industry*, vol. 2, *1700–1830: The Industrial Revolution* [Oxford: Clarendon Press, 1984], 26; Roy Church, *The History of the British Coal Industry*, vol. 3, *1830–1913: Victorian Pre-eminence* [Oxford: Clarendon Press, 1986], 3.)

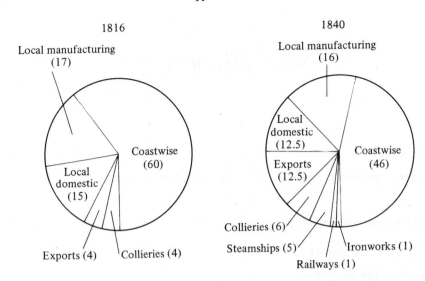

Figure A.2. Estimated percent consumption of coal output: Durham and Northumberland, 1816 and 1840. (From B. R. Mitchell, *Economic Development of the British Coal Industry, 1800–1914* [Cambridge University Press, 1984], 16–17.)

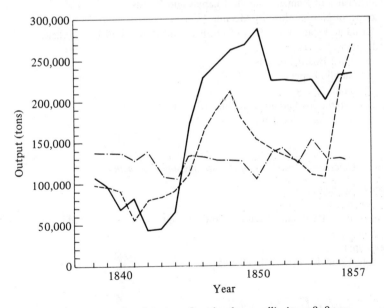

Figure A.3. Distribution of coal output: Londonderry collieries, 1838–57. Key: ---, Small; —, E. Main; -·-, Wallsend. (From D.C.R.O., Londonderry Papers, D/Lo/E 514(8).)

Select bibliography

Place of publication is London unless otherwise noted.

Manuscript and typescript sources

Durham County Record Office

Durham Quarter Sessions
  Order Books
  Special County Rate for Gaol, 1809 and 1819
  Militia Family Relief Payments, 1807–11
  Durham Recusants' Estates, 1717–1778
  Certificates of Roman Catholic Chapels and Priests
  Return of Non-conformist Meeting Houses
  Charitable Estate of the City of Durham and Borough of Framwelgate
Londonderry Papers
National Coal Board Records
  John Buddle Papers
  T. Y. Hall Papers
Primitive Methodist Circuit Accounts, 1828–1836
Speculum Gregis, Whickham

Durham University, Department of Paleography and Diplomatic

Diocese Book, 1793
Durham Bishopric Rentals
Records of the Palatinate of Durham and Bishopric Estates deposited by the Church
  Commission
Grey Papers
Halmote Court Records, Miscellanea

Lambton Estate Archives, Chester-le-Street

Lambton Mss.

Newcastle Central Library, Newcastle-upon-Tyne

Matthias Dunn's Diary

North of England Institute of Mining and Mechanical Engineers, Newcastle-upon-Tyne

Bell Collection
Watson Collection

Northumberland Record Office, Newcastle-upon-Tyne

Delaval Papers
Minutes of the Committee Meetings of the Coal Owners of the Rivers Tyne and
      Wear
Minutes of the General Meetings of the Coal Owners of the Rivers Tyne and Wear

Public Record Office, London and Kew

Home Office Papers
Land Tax Assessments
Durham Assize and Gaol Delivery

Public Record Office of Northern Ireland, Belfast

Londonderry Papers

Tyne and Wear County Archives, Newcastle-upon-Tyne

Houghton-le-Spring Circuit, Shiney Row Chapel, Missionary Society, 1826–1844
Newcastle and Brunswick Central Circuit, Brunswick Methodist Chapel, Register
      and Minute Book, 1825–1835
South Shields Circuit, Members, 1811–1840
South Shields Circuit, Members, 1832–1841
South Shields Primitive Methodist Chapel, Accounts of Class Money, 1831–1840

Periodicals

Durham Chronicle
Gateshead Observer
Newcastle Chronicle
Newcastle Courant
Northern Star
Tyne Mercury

Parliamentary papers

Accounts and papers

Abstract of Answers and Returns, made pursuant to an Act 2 Geo. 4, for taking an Account of the Population of Great Britain, and of the Increase and Diminution thereof, 1833, vols. 36–8.

Abstract of Answers and Returns under the Population Act, 3 & 4 Vic., c. 99, 1843, vol. 22.

Account: Population and Houses, 1841, 1841, vol. 2.

Population (Great Britain): Religious Worship (England and Wales), 1852–53, vol. 89.

Reports from commissioners

Reports from Commissioners: Poor Laws: Appendix A, Part 1, 1834, vol. 28.

Reports from Commissioners: Children's Employment (Mines), 1842, vol. 16.

Reports from Commissioners: Population (Age Abstract), 1843, vol 23.

Reports from Commissioners: Population: Occupation Abstract, 1844, vol. 27.

Report of the Commissioner appointed under the Provisions of the Act 5 & 6 Vict., c. 99, to Inquire into the Operation of the Act, and into the State of the Population of the Mining Districts, 1846, 1846, vol. 24.

Report of the Commissioners appointed to enquire into the Several Matters relating to Coal in the United Kingdom, 1871, c. 435, vol. 28, 1; C. 435–1, vol. 18, 199; C. 435–2, vol. 18, 815.

Reports from committees

Report from the Committee Appointed to Consider the Coal Trade of This Kingdom, 1800, First Series: vol. 10, 538.

Report from the Select Committee Appointed to Enquire into the State of the Coal Trade of this Kingdom, 1800, First Series: vol. 10, 640.

Report from the Select Committee of the House of Lords Appointed to Take into Consideration the State of the Coal Trade in the United Kingdom, 1830 (9), vol. 8, 405.

Report from the Select Committee Appointed to Inquire into the State of the Coal Trade in the Port of London, 1830 (663), vol. 8, 1.

Report from the Select Committee appointed to Inquire into the Nature, Cause and Extent of the Lamentable Catastrophes which have occurred in the Mines of Great Britain, 1835 (603), vol. 5, 1.

Report from the Select Committee Appointed to Inquire into the State of the Coal Trade, as respects the Supply of Coal into the Port of London, and the Adjacent Counties, from the Rivers Tyne, Wear, and Tees, and Other Places, 1836 (522), vol. 11, 169.

Report from the Select Committee on Church Leases, 1837–8 (692), vol. 9, 1.

Unpublished theses and dissertations

Cromar, Peter, "Economic Power and Organization: The Development of the Coal Industry on Tyneside, 1700–1828," Ph.D. dissertation, University of Cambridge (1977).

Hair, P. E. H., "The Social History of the British Coal Miners, 1800–1845," D. Phil. dissertation, University of Oxford (1955).

Johnson, W. H., "A North-East Miners' Union (Hepburn's Union) of 1831–2," M.A. thesis, University of Durham (1959).

Books and pamphlets

An Appeal to the Public, from the Pitmen of the Tyne and Wear (Newcastle-upon-Tyne, 1832; repr. New York: Arno, 1972).

Bailey, John, *General View of the Agriculture of the County of Durham* (1810).

A Candid Appeal to the Coal Owners and Viewers of the Collieries on the Tyne and Wear (Newcastle-upon-Tyne, 1826).

Castor, *A Letter on the Disputes between the Coal-Owners & Pitmen, addressed to the Tyne Mercury* (Newcastle-upon-Tyne, 1832; repr. New York: Arno, 1972).

Cobbett, William, *A Tour in Scotland and the Four Northern Counties* (1833).

Dunn, Matthias, *An Historical, Geological and Descriptive View of the Coal Trade of the North of England* (Newcastle-upon-Tyne, 1844).

Durham, J. G. Lambton, First Earl, *The Revival of the Right Principles in the North* (1833).

Speeches of the Earl of Durham on the Reform of Parliament (1835).

[Errington, Anthony], *Coals on Rails: Or the Reason of My Wrighting*, P. E. H. Hair, ed. (Liverpool: Liverpool University Press, 1988).

Extracts from the Records of the Company of Hostmen of Newcastle-upon-Tyne (Newcastle-upon-Tyne: Surtees Society Publication no. 150, 1901).

Fordyce, William, *The History and Antiquities of the County Palatine of Durham*, 2 vols. (Newcastle-upon-Tyne, 1857).

A Freeman, *The Contest, being a complete collection of the Controversial Papers including Poems and Songs during the Late Contested Election for the City of Durham in March, 1800* (Durham, 1800).

Granger, Joseph, *General View of the Agriculture of the County of Durham* (1794).

Hodgskin, Thomas, *Labour Defended Against the Claims of Capital* (1825; repr. Labour Publishing, 1922).

Popular Political Economy: Four Lectures delivered at the London Mechanics' Institution (1827; repr. New York: Kelley, 1966).

Holland, J., *The History and Description of Fossil Fuel, the Collieries, and Coal Trade of Great Britain* (1841).

Hutchinson, William, *The History and Antiquities of the County Palatine of Durham*, 3 vols. (Newcastle-upon-Tyne, 1787).

The Large Hymnbook for the Use of Primitive Methodists (Bemersley, near Tunstall, 1825).

Leifchild, J. R., *Our Coal and our Coal-Pits* (1856; repr. New York: Kelley, 1968).
Losh, James, *The Diaries and Correspondence of James Losh, 1811–1833,* Edward Hughes, ed., 2 vols. (Newcastle-upon-Tyne: Surtees Society Publications nos. 171 and 174, 1962–3).
Mitchell, William, *The Question Answered: "What do the Pitmen Want?"* (Bishopwearmouth, 1844; repr. New York: Arno, 1972).
Parson, William, and White, William, *History, Directory and Gazetteer of the Counties of Durham and Northumberland,* 2 vols. (Leeds, 1827–8).
Report by the Committee of Coalowners respecting the Present Situation of the Trade (Newcastle-upon-Tyne, 1832; repr. New York: Arno, 1972).
Report of Speeches delivered at a Public Meeting of the Friends of the Established Church (Newcastle-upon-Tyne, 1834).
Report of the Trials of the Pitmen and Others, concerned in the Late Riots, Murders, &c., in the Hetton and other Collieries at the Durham Summer Assizes, 1832 (Durham, 1832; repr. New York: Arno, 1972).
Richardson, M. A., *The Local Historian's Table Book,* 5 vols. (1843).
Rules and Regulations for the Formation of a Society to be Called the United Association of Colliers of the Rivers Tyne and Wear (Newcastle-upon-Tyne, 1825).
Rules and Regulations of the Coal Miners' Friendly Society in the Counties of Northumberland and Durham (Newcastle, 1832).
Rymer, E. A., *The Martyrdom of the Mine* (1898; repr. *History Workshop,* Robert G. Neville, ed., vols. 1–2, Spring–Autumn, 1976).
Scott, William, *An Earnest Address, and Urgent Appeal to the People of England, in Behalf of the Oppressed and Suffering Pitmen, of the Counties of Northumberland and Durham* (Newcastle-upon-Tyne, 1831; repr. New York: Arno, 1972).
Scrutator, *An Impartial Enquiry into the Existing Causes of Dispute between the Coal Owners of the Wear and Tyne, and their Late Pitmen* (Houghton-le-Spring, 1832; repr. New York: Arno, 1972).
Thompson, William, *An Inquiry into the Principles of the Distribution of Wealth Most Conducive to Human Happiness* (1824; repr. New York: Kelley, 1963).
[Thompson, William] *Labour Rewarded: The Claims of Labour and Capital Conciliated, or How to Secure to Labour the Whole Product of Its Exertion* (1827; repr. New York: Kelley, 1969).
A Voice from the Coal Mines; or, A Plain Statement of the Various Grievances of the Pitmen of the Tyne and Wear (South Shields, 1825).
Whellan, Francis, *History, Topography and Directory of the County Palatine of Durham* (1894).

Secondary literature

Books

Andrews, William, ed., *Bygone Durham* (W. Andrews & Co., 1898).
Ashton, T. S. and Sykes, J., *The Coal Industry of the Eighteenth Century* (Manchester: Manchester University Press, 1929).

Atkinson, Frank, *The Great Northern Coalfield, 1700–1900* (Barnard Castle: Durham County Local History Society, 1966).

Bain, Joe S., *Barriers to New Competition* (Cambridge, Mass.: Harvard University Press, 1956).

Barker, T. C., and Harris, J. R., *A Merseyside Town in the Industrial Revolution: St. Helens, 1750–1900* (Liverpool: Liverpool University Press, 1954).

Baugh, D. A., ed., *Aristocratic Government and Society in Eighteenth Century England* (New York: New Viewpoints, 1975).

Bean, W. W., *The Parliamentary Representation of the Six Northern Counties of England* (Hull: Charles Henry Barnwell, 1890).

Bebbington, D. W., *Evangelicalism in Modern Britain: A History of the 1730s to the 1980s* (Unwin Hyman, 1989).

Benson, John, *British Coalminers in the Nineteenth Century: A Social History* (Dublin: Gill & Macmillan, 1980).

Benson, John, Neville, R. G., and Thompson, C. H., *Bibliography of the British Coal Industry: Secondary Literature, Parliamentary and Departmental Papers, Mineral Maps and Plans and a Guide to Sources* (Oxford: Oxford University Press, 1981).

Berg, Maxine, *The Age of Manufactures, 1700–1820* (Fontana, 1985).

Berg, Maxine, Hudson, Pat, and Sonenscher, Michael, eds., *Manufacture in Town and Country Before the Factory* (Cambridge: Cambridge University Press, 1983).

Best, G. F. A., *Temporal Pillars* (Cambridge: Cambridge University Press, 1964).

Bossy, John, *The English Catholic Community, 1570–1850* (Darton, Longman & Todd, 1975).

Bottomore, Tom, *Theories of Modern Capitalism* (Allen & Unwin, 1985).

Bovil, E. M., *The England of Nimrod and Surtees* (Oxford: Oxford University Press, 1959).

Bulmer, Martin, ed., *Mining and Social Change: Durham County in the Twentieth Century* (Croom Helm, 1978).

Burawoy, Michael, *Manufacturing Consent: Changes in the Labour Process Under Monopoly Capitalism* (Chicago. University of Chicago Press, 1979).

The Politics of Production: Factory Regimes under Capitalism and Socialism (Verso, 1985).

Burgess, Keith, *The Origins of British Industrial Relations* (Croom Helm, 1975).

Buxton, Neil, *The Economic Development of the British Coal Industry* (Batsford, 1978).

Cadogan, Peter, *Early Radical Newcastle* (Consett: Sagittarius Press, 1975).

Calhoun, Craig, *The Question of Class Struggle: Social Foundations of Popular Radicalism during the Industrial Revolution* (Chicago: University of Chicago Press, 1982).

Cameron, Rondo, ed., *Banking and Economic Development* (New York: Oxford University Press, 1972).

Cameron, Rondo, with Crisp, Olga, Patrick, Hugh T., and Tilly, Richard, *Banking in the Early Stages of Industrialization: A Study in Comparative Economic History* (New York: Oxford University Press, 1967).

Challinor, Raymond, and Ripley, Brian, *The Miners' Association: A Trade Union in the Age of the Chartists* (Lawrence & Wishart, 1968).

Chamberlin, Edward, *The Theory of Monopolistic Competition* (Cambridge, Mass.: Harvard University Press, 1939).

Chambers, J. D., and Mingay, G. E., *The Agricultural Revolution, 1750–1880* (Batsford, 1966).

Chandler, Alfred, *The Visible Hand: The Mangerial Revolution in American Business* (Cambridge, Mass.: Harvard University Press, 1977).

Chapman, S. D., *The Cotton Industry in the Industrial Revolution* (Macmillan, 1972).

Church, Roy, and Chapman, S. D., "Gravener Henson and the Making of the English Working Class," in E. L. Jones and G. E. Mingay, eds., *Land, Labour and Population in the Industrial Revolution* (Edward Arnold, 1967), 131–61.

Church, Roy, with Hall, Alan, and Kanefsky, John, *The History of the British Coal Industry, Volume 2: 1830–1913: Victorian Pre-eminence* (Oxford: Clarendon, 1986).

Claeys, Gregory, *Machinery, Money and the Millenium: From Moral Economy to Socialism, 1815–1860* (Princeton: Princeton University Press, 1987).

Citizens and Saints: Politics and Anti-politics in Early British Socialism (Cambridge: Cambridge University Press, 1989).

Clarke, John, Critcher, Chas, and Johnson, Richard, eds., *Working Class Culture: Studies in History and Theory* (Hutchinson, 1979).

Clegg, H. A., Fox, Alan, and Thompson, A. F., *A History of British Trade Unions since 1889*, vol. 1, 1889–1910 (Oxford: Clarendon, 1964).

Colls, Robert, *The Collier's Rant* (Croom Helm, 1977).

The Pitmen of the Northern Coalfield: Work, Culture, and Protest, 1790–1850 (Manchester: Manchester University Press, 1987).

Corbin, David A., *Life, Work and Rebellion in the Coal Fields: The Southern West Virginia Miners, 1880–1922* (Chicago: University of Illinois Press, 1981).

Cousins, J. M., and Davis, R. L., " 'Working Class Incorporation'–A Historical Approach with Reference to the Mining Communities of S.E. Northumberland, 1840–1890," in Frank Parkin, ed., *The Social Analysis of Class Structure* (Tavistock, 1974), 275–97.

Crafts, N. F. R., *British Economic Growth during the Industrial Revolution* (Oxford: Clarendon, 1985).

Creamer, D., Dobrovolsky, S. P., and Borenstein, I., *Capital in Manufacturing and Mining: Its Formation and Financing* (Princeton: Princeton University Press, 1960).

Crew, David, *Town in the Ruhr: A Social History of Bochum, 1860–1914* (New York: Columbia University Press, 1979).

Crouzet, François, ed., *Capital Formation in the Industrial Revolution* (Methuen & Co., 1972).

The First Industrialists: The Problem of Origins (Cambridge: Cambridge University Press, 1985).

Currie, Robert, *Methodism Divided: A Study in the Sociology of Ecumenicalism* (Faber, 1968).

Currie, R., Gilbert, A., and Horsley, L., *Churches and Churchgoers: Patterns of Church Growth in the British Isles since 1700* (Oxford: Clarendon, 1977).

Deane, Phyllis, *The First Industrial Revolution*, 2nd ed. (Cambridge: Cambridge University Press, 1979).

Dennis, Norman, Henriques, Fernando, and Slaughter, Clifford, *Coal is Our Life* (1956; repr. Tavistock, 1969).

Douglass, David, and Krieger, Joel, *A Miner's Life* (Routledge & Kegan Paul, 1983).

Duckham, Baron F. and Duckham, Helen, *Great Pit Disasters* (Newton Abbot: David & Charles, 1973).

Eastwood, T., *British Regional Geology: Northern England*, 2nd ed. (HMSO, 1946).

Edwards, P. K., *Conflict at Work: A Materialist Analysis of Workplace Relations* (Oxford: Blackwell Publisher, 1986).

Elbaum, Bernard, and Lazonick, William, eds., *The Decline of the British Economy* (Oxford: Clarendon, 1986).

Epstein, James, and Thompson, Dorothy, eds., *The Chartist Experience: Studies in Working-Class Radicalism and Culture, 1830–1860* (Macmillan, 1982).

Evans, Neil, "Two Paths to Economic Development: Wales and the North-East of England," in Pat Hudson, ed., *Regions and Industries: A Perspective on the Industrial Revolution in Britain* (Cambridge: Cambridge University Press, 1989), 201–27.

Everitt, Alan, *The Pattern of Rural Dissent: The Nineteenth Century* (Leicester: Leicester University Press, 1972).

Flanders, Allan, *Industrial Relations: What is Wrong with the System?* (Faber & Faber, 1965).

Flinn, Michael W., with Stoker, David, *The History of the British Coal Industry, Volume 2: 1700–1830: The Industrial Revolution* (Oxford: Clarendon, 1984).

Floud, Roderick, and McCloskey, Donald, eds., *The Economic History of Britain since 1700*, Vol. 1, 1700–1860 (Cambridge: Cambridge University Press, 1981).

Fontana, Biancamaria, *Rethinking the Politics of Commercial Society: The Edinburgh Review, 1802–1832* (Cambridge: Cambridge University Press, 1985).

Foster, John, *Class Struggle and the Industrial Revolution: Early Industrial Capitalism in Three English Towns* (Weidenfeld & Nicolson, 1974).

Fynes, Richard, *The Miners of Northumberland and Durham* (1873; repr. S. R. Publishers Ltd., 1971).

Galloway, Robert L., *A History of Coal Mining in Great Britain* (1882; repr. Newton Abbot: David & Charles, 1969).

Annals of Coal Mining and the Coal Trade, 2 vols. (1898; repr. Newton Abbot: David & Charles, 1971).

Garside, W. R., *The Durham Miners, 1919–1960* (Allen & Unwin, 1971).

Gayer, A., Rostow, W. W., and Schwartz, A. J., *The Growth and Fluctuation of the British Economy, 1790–1850*, 2 vols. (Oxford: Clarendon, 1953).

Geiger, Reed G., *The Anzin Coal Company, 1800–1833* (Newark: University of Delaware Press, 1974).

Gilbert, A. D., *Religion and Society in Industrial England: Church, Chapel and Social Change, 1740–1914* (Longman Group, 1976).

Glen, Robert, *Urban Workers in the Early Industrial Revolution* (Croom Helm, 1984).

Goodrich, Carter, *The Frontier of Control: A Study in British Workshop Politics* (1920; repr. Pluto Press, 1975).

The Miner's Freedom: A Study of the Working Life in a Changing Industry (Boston: Marshall Jones, 1925).

Gospel, Howard F., and Littler, Craig R., eds., *Managerial Strategies and Industrial Relations* (Heinemann, 1983).

Gray, Robert Q., *The Labour Aristocracy in Victorian Edinburgh* (Oxford: Clarendon, 1976).

Haines, Michael, *Fertility and Occupation: Population Patterns in Industrialization* (New York: Academic, 1979).

Halévy, Elie, *A History of the English People in 1815* (1924; repr. Ark ed. 1987).

Hammond, J. L., and Hammond, Barbara, *The Skilled Labourer* (1919; repr. Longman Group, 1979).

Harrison, Brian, and Hollis, Patricia, eds., *Robert Lowery: Radical and Chartist* (1979).

Harrison, Royden, ed., *Independent Collier: The Coal Miner as Archetypal Proletarian Reconsidered* (New York: St. Martin's, 1978).

Harrison, Royden, and Zeitlin, Jonathan, eds., *Divisions of Labour: Skilled Workers and Technological Change in Nineteenth Century England* (Sussex: Harvester, 1985).

Hawke, G. R., *Railways and Economic Growth in England and Wales, 1840–1870* (Oxford: Oxford University Press, 1970).

Hempton, David, *Methodism and Politics in British Society, 1750–1850* (Hutchinson, 1984).

Higgins, J. P. P. and Pollard, Sidney, eds., *Aspects of Capital Investment in Great Britain, 1750–1850* (Methuen & Co., 1971).

Hinton, James, *Labour and Socialism: A History of the British Labour Movement, 1867–1974* (Brighton: Wheatsheaf, 1983).

Hobsbawm, E. J., *Primitive Rebels* (New York: Norton, 1959).

The Age of Revolution, 1798–1848 (New York: New American Library, 1962).

Labouring Men: Studies in the History of Labour (Weidenfeld & Nicolson, 1964).

Industry and Empire (Harmondsworth: Penguin Books, 1968).

Worlds of Labor (Weidenfeld & Nicolson, 1984).

Hobsbawm, E. J. and Rudé, George, *Captain Swing* (1969; repr. Harmondsworth: Penguin Books, 1973).

House, J. W., *The North-East* (Newton Abbot: David & Charles, 1969).

Hudson, Pat, *The Genesis of Industrial Capital: A Study of the West Riding Wool Textile Industry, c. 1750–1850* (Cambridge: Cambridge University Press, 1986).

Hughes, Edward, *North Country Life in the Eighteenth Century: The North-East, 1700–1750* (1952; repr. Oxford: Oxford University Press, 1969).

"The Professions in the Eighteenth Century," repr. in D. A. Baugh, ed., *Aristocratic Government and Society in Eighteenth Century England* (New York: New Viewpoints, 1975).

James, Mervyn, *Family, Lineage, and Civil Society: A Study of Society, Politics, and Mentality in the Durham Region, 1500–1640* (Oxford: Clarendon, 1974).

John, A. H., *The Industrial Development of South Wales, 1750–1850* (Cardiff: University of Wales Press, 1950).

Jones, E. L., *Seasons and Prices* (Allen & Unwin, 1964).

Jones, E. L., and Mingay, G. E. (eds.), *Land, Labour and Population in the Industrial Revolution* (Arnold, 1967).

Joyce, Patrick, *Work, Society and Politics: The Culture of the Factory in Later Victorian England* (Brighton: Harvester, 1980).

Joyce, Patrick, ed., *The Historical Meanings of Work* (Cambridge: Cambridge University Press, 1987).

Kellett, J. R., *The Impact of Railways on Victorian Cities* (Routledge & Kegan Paul, 1969).

Kendall, H. B., *The Origin and History of the Primitive Methodist Church*, 2 vols. (Dalton, 1909).

Kent, John, *Holding the Fort: Studies in Victorian Revivalism* (Epworth, 1978).

Kenwood, A. G., *Capital Formation in North East England, 1800–1913* (New York: Garland, 1985).

Kirby, R. G. and Musson, A. E., *The Voice of the People: John Doherty, 1798–1854, Trade Unionist, Radical and Factory Reformer* (Manchester: Manchester University Press, 1975).

Kirk, Neville, *The Growth of Working Class Reformism in Mid-Victorian England* (Croom Helm, 1985).

Köllman, Wolfgang, *Bevölkerung in der industriellen Revolution, Kritische Studien zur Geschictswissenschaft*, bd. 12 (Göttingen: Vandenhoeck & Ruprecht, 1974).

Landes, David, *The Unbound Prometheus: Technological Change and Industrial Development in Western Europe from 1750 to the Present* (Cambridge: Cambridge University Press, 1969).

Langton, John, *Geographical Change and the Industrial Revolution: Coal-Mining in South-West Lancashire, 1590–1799* (Cambridge: Cambridge University Press, 1979).

Lapsely, Gaillard, *The County Palatine of Durham* (Longman Group, 1900).

Laqueur, Thomas W., *Religion and Respectability: Sunday Schools and Working Class Culture, 1780–1850* (New Haven: Yale University Press, 1976).

Lawson, Jack, *A Man's Life* (Hodder & Stoughton, 1949).

Levy, Hermann, *Monopolies, Cartels and Trusts in British Industry* (1909; repr. Cass, 1968).

Lewis, Brian, *Coal Mining in the Eighteenth and Nineteenth Centuries* (Longman Group, 1971).

Lewis, M. J. T., *Early Wooden Railways* (Routledge & Kegan Paul, 1970).

Mathias, Peter, *The Brewing Industry in England, 1700–1830* (Cambridge: Cambridge University Press, 1959).

The First Industrial Nation: An Economic History of Britain, 1700–1814 (New York: Scribner, 1969).

The Transformation of England: Essays in the Economic and Social History of England in the Eighteenth Century (Methuen & Co., 1979).

Mathias, Peter and Postan, M. M., eds., The Cambridge Economic History of Europe, vol. 7, *The Industrial Economies: Capital, Labour, and Enterprise*, pt. 1 (Cambridge: Cambridge University Press, 1978).

Matsumura, Takao, *The Labour Aristocracy Revisited: The Victorian Flint Glass Makers, 1850–1880* (Manchester: Manchester University Press, 1983).

Matthews, R. C. O., *A Study in Trade Cycle History* (Cambridge: Cambridge University Press, 1954).

McLeod, Hugh, *Class and Religion in the Late Victorian City* (Croom Helm, 1974).

Religion and the Working Class in Nineteenth-Century Britain (Macmillan Press, 1984).

Melling, Joseph, "Employers, Industrial Welfare, and the Struggle for Work-place Control in British Industry, 1880–1920," in Howard F. Gospel and Craig R. Littler, eds., *Managerial Strategies and Industrial Relations* (Heinemann, 1983).

Mitchell, B. R., *Economic Development of the British Coal Industry, 1800–1914* (Cambridge: Cambridge University Press, 1984).

Moore, Robert, *Pit-Men, Preachers and Politics: The Affects of Methodism in a Durham Mining Community* (Cambridge: Cambridge University Press, 1974).

Morris, R. J., *Cholera 1832* (Croom Helm, 1976).

Class and Class Consciousness in the Industrial Revolution, 1780–1850 (Macmillan, 1979).

Nef, J. U., *The Rise of the British Coal Industry*, 2 vols. (G. Routledge & Sons, 1932).

New, Chester W., *Lord Durham: A Biography of John George Lambton, First Earl of Durham* (Oxford: Oxford University Press, 1929).

Nossiter, T. J., *Influence, Opinion and Political Idioms in Reformed England: Case Studies from the North-East, 1832–1874* (Hassocks: Harvester, 1974).

Obelkevich, James, *Religion and Rural Society: South Lindsey, 1825–1875* (Oxford: Oxford University Press, 1976).

Oxberry, John, *Thomas Hepburn of Felling: What He Did for the Miners* (Felling-on-Tyne: Heslop, 1959).

Parkin, Frank, ed., *The Social Analysis of Class Structure* (Tavistock, 1974).

Patterson, William, *Northern Primitive Methodism* (E. Dalton, 1909).

Pelling, Henry, *A History of British Trade Unionism* (Harmondsworth: Penguin Books, 1963).

Perkin, Harold, *The Origins of Modern English Society, 1780–1880* (Routledge & Kegan Paul, 1969).

Petty, John, *The History of the Primitive Methodist Connexion* (1860).

Pollard, Sidney, *The Genesis of Modern Management* (Cambridge, Mass.: Harvard University Press, 1965).

Pressnell, L. S., *Country Banking in the Industrial Revolution* (Oxford: Oxford University Press, 1956).

(ed.), *Studies in the Industrial Revolution: Presented to T. S. Ashton* (Athlone, 1960).

Price, Richard, *Masters, Unions and Men: Work Control in Building and the Rise of Labour, 1830–1914* (Cambridge: Cambridge University Press, 1980).

Labour in British Society (1986; repr. Routledge, 1990).

Prothero, Iorwerth, *Artisans and Politics in Early Nineteenth-Century London: John Gast and His Times* (Dawson, 1979).

Raybould, T. J., *The Economic Emergence of the Black Country: A Study of the Dudley Estate* (Newton Abbot: David & Charles, 1973).

Reddy, William M., *The Rise of Market Culture: The Textile Trade and French Society, 1750–1900* (Cambridge: Cambridge University Press, 1984).

Redford, Arthur, *Labour Migration in England, 1800–1850* (1926; 3rd ed., Manchester: Manchester University Press, 1976).

Reid, Stuart J., *Life and Letters of the First Earl of Durham, 1792–1840*, 2 vols. (Longman Group, 1906).

Robinson, Joan, *The Economics of Imperfect Competition*, 2nd ed. (Macmillan Press, 1969).

Roemer, John E., *A General Theory of Exploitation and Class* (Cambridge, Mass.: Harvard University Press, 1982).

Rowe, J. W. F., *Wages in the Coal Industry* (P. S. King & Son, 1923).

Rudé, George, *Ideology and Popular Protest* (New York: Pantheon, 1980).

Rule, John, *The Labouring Classes in Early Industrial England 1750–1850* (Longman Group, 1986).

Samuel, Raphael, ed., *Miners, Quarrymen and Saltworkers* (Routledge & Kegan Paul, 1977).

Schluchter, Wolfgang, *The Rise of Western Rationalism: Max Weber's Developmental History*, G. Roth, trans. (Berkeley and Los Angeles, University of California Press, 1981).

Schubert, Adrian, "Private Initiative in Law Enforcement: Associations for the Prosecution of Felons, 1744–1856," in Victor Bailey, ed., *Policing and Punishment in Nineteenth-Century Britain* (New Brunswick, N.J.: Rutgers University Press, 1981).

Schumpeter, Joseph A., *Business Cycles: A Theoretical, Historical and Statistical Analysis of the Capitalist Process*, 2 vols. (New York: McGraw-Hill, 1939).

Smith, Raymond, *Sea Coal for London: History of the Coal Factors in the London Market* (Longman Group, 1961).

Sonenscher, Michael, *The Hatters of Eighteenth-Century France* (Berkeley and Los Angeles: University of California Press, 1987).

Work and Wages: Natural Law, Politics and the Eighteenth-Century French Trades (Cambridge: Cambridge University Press, 1989).

Spring, David, *The English Landed Estate in the Nineteenth Century: Its Administration* (Baltimore: Johns Hopkins University Press, 1963).

Stedman Jones, Gareth, *Languages of Class: Studies in English Working Class History, 1832–1982* (Cambridge: Cambridge University Press, 1983).

Sturgess, R. W., *Aristocrat In Business: The Third Marquess of Londonderry as Coalowner and Portbuilder* (Durham County Local History Society, 1975).

Sweezy, Paul M., *Monopoly and Competition in the English Coal Trade, 1550–1850* (Cambridge, Mass.: Harvard University Press, 1938).

Sylos Labini, Paolo, *Oligopoly and Technical Progress* (rev. ed., Cambridge, Mass.: Harvard University Press, 1969).

Taylor, A. J., "The Sub-contract System in the British Coal Industry," in L. S. Pressnell, ed., *Studies in the Industrial Revolution* (Athlone, 1960), 215–35.

Thompson, David M., *Nonconformity in the Nineteenth Century* (Routledge & Kegan Paul, 1972).

Thompson, E. P., *The Making of the English Working Class* (New York: Vintage, 1963).

Thompson, F. M. L., *English Landed Society in the Nineteenth Century* (Routledge & Kegan Paul, 1963).

Thompson, Noel W., *The People's Science: The Popular Political Economy of Exploitation and Crisis, 1816–1834* (Cambridge: Cambridge University Press, 1984).

The Market and Its Critics: Socialist Political Economy in Nineteenth Century Britain (Routledge, 1988).

Thompson, Paul, *The Nature of Work: An Introduction to Debates on the Labour Process* (Macmillan Press, 1983).

von Tunzelmann, G. N., *Steam Power and British Industrialization to 1860* (Oxford: Clarendon, 1978).

Valenze, Deborah, *Prophetic Sons and Daughters: Female Preaching and Popular Religion in Industrial England* (Princeton: Princeton University Press, 1985).

Ward, J. T. and Wilson, R. G., *Land and Industry: The Landed Estate and the Industrial Revolution* (New York: Barnes & Noble, 1971).

Ward, W. R., *Religion and Society in England, 1790–1850* (Batsford, 1972).

Watson, Aaron, *A Great Labour Leader: Being the Life of the Right Honourable Thomas Burt* (Brown, Langham, 1908).

Webb, Sidney, *The Story of the Durham Miners, 1662–1921* (Fabian, 1921).

Webb, Sidney, and Webb, Beatrice, *The History of Trade Unionism*, new ed. (Longman Group, 1911).

Weber, Max, *Economy and Society*, 2 vols., G. Roth and C. Wittich, eds. (1956., 4th ed., repr. Berkeley and Los Angeles: University of California Press, 1978).

Welbourne, E., *The Miners' Unions of Northumberland and Durham* (Cambridge: Cambridge University Press, 1923).

Werner, J. S., *The Primitive Methodist Connexion* (Madison: University of Wisconsin Press, 1984).

Wiener, Martin, *English Culture and the Decline of the Industrial Spirit, 1850–1980* (Cambridge: Cambridge University Press, 1981).

Wilcock, Don, *The Durham Coalfield*, Pt. 1, *The "Sea Cole" Age* (Durham: Durham County Council, 1979).

Willan, T. S., *The English Coasting Trade, 1600–1750* (Manchester: Manchester University Press, 1938).

Wilson, John, *A History of the Durham Miners' Association, 1870–1904* (Durham: Veitch, 1907).

Memories of a Labour Leader: The Autobiography of John Wilson, J.P., M.P. (Unwin, 1910).

Wright, Erik Olin, *Classes* (Verso, 1985).

Articles

Amsden, Jon and Briar, Stephen, "Coal Miners on Strike: The Transformation of Strike Demands and the Formation of a National Union," *Journal of Interdisciplinary History*, 7, no. 4 (1977), 583–616.

Bulmer, M. I. A., "Sociological Models of the Mining Community," *Sociological Review*, n.s., 23, no. 1, (1975), 61–92.

Charlesworth, Andrew, and Randall, Adrian J., "Comment: Morals, Markets and the English Crowd in 1766," *Past & Present*, no. 114 (February 1987), 200–213.

Colls, Robert, " 'Oh Happy English Children!': Coal, Class and Education in the North-East," *Past & Present*, no. 73 (November 1976), 75–99.

"Debate: Coal, Class and Education in the North-East: A Rejoinder," *Past & Present*, no. 90 (February 1981), 152–65.

Corrigan, J. V., "Strikes and the Press in the North-East, 1815–1844: A Note," *International Review of Social History*, 23 (1978), 376–81.

Crafts, N. F. R., "Average Age at First Marriage for Women in Mid-Nineteenth

Century England and Wales: A Cross-Section Study," *Population Studies*, 33 no. 1 (1978), 21–5.

Cromar, Peter, "The Coal Industry on Tyneside, 1771–1800: Oligopoly and Spatial Change," *Economic Geography*, 53, no. 1 (1977), 79–94.

"The Coal Industry on Tyneside, 1715–1750," *Northern History*, 14 (1978), 193–207.

Currie, Robert, "A Micro-Theory of Methodist Growth," *Proceedings of the Wesley Historical Society*, 36 (October 1967), 65–73.

Daunton, M. J., "Miners' Houses: South Wales and the Great Northern Coalfield, 1880–1914," *International Review of Social History*, 25 (1980), 143–75.

"Down the Pit: Work in the Great Northern and South Wales Coalfields, 1870–1914," *Economic History Review*, 2nd ser., 34, no. 4 (1981), 578–97.

Duffy, Brendan, "Debate; Coal, Class and Education in the North-East," *Past & Present*, no. 90 (February 1981), 142–51.

Dutton, H. I., and King, J. E., "The Limits of Paternalism: The Cotton Tyrants of North Lancashire, 1836–54," *Social History*, 7, no. 1 (1982), 59–74.

Fewster, J. M., "The Keelmen of Tyneside in the Eighteenth Century," pts. 1–3, *Durham University Journal*, n.s., 19, no. 1, 24–33; 19, no. 2, 66–75; 19, no. 3, 111–123 (1957–8).

Friedlander, Dov, "Demographic Patterns and Socioeconomic Characteristics of the Coal-mining Population in England and Wales in the Nineteenth Century," *Economic Development and Cultural Change*, 22, no. 1 (1973), 39–51.

Friedman, Andy, "Responsible Autonomy versus Direct Control over the Labour Process," *Capital & Class*, no. 1 (Spring 1977), 43–57.

Gilbert, A. D., "Methodism, Dissent and Political Stability in Early Industrial England," *Journal of Religious History*, 10, no. 4 (1979), 381–99.

Griffin, A. R., "Methodism and Trade Unionism in the Nottinghamshire-Derbyshire Coalfield, 1844–1890," *Proceedings of the Wesley Historical Society*, 37, no. 2 (1969), 2–9.

Hair, P. E. H., "The Binding of the Pitmen of the North-East, 1800–1809," *Durham University Journal*, n.s., 27, no. 3 (1965), 1–13.

"Mortality from Violence in British Coal-Mines, 1800–1850," *Economic History Review*, 2nd ser., 21, no. 3 (1968), 545–61.

Harris, J. R., "Skills, Coal and British Industry in the Eighteenth Century," *History*, 61, no. 202 (1976), 167–82.

Harrison, Brian, "Class and Gender in Modern British Labour History," *Past & Present*, no. 124 (August 1989), 121–58.

Hausman, William J., "Size and Profitability of English Colliers in the Eighteenth Century," *Business History Review*, 51 (1977), 460–73.

"The English Coastal Coal Trade, 1691–1910: How Rapid Was Productivity Growth?" *Economic History Review*, 2nd ser., 40, no. 4 (1987), 588–96.

Heesom, A. J., "Entrepreneurial Paternalism: The Third Lord Londonderry (1778–1854) and the Coal Trade," *Durham University Journal*, n.s., 35, no. 3 (1974), 238–56.

"Problems of Patronage: Lord Londonderry's Appointment as Lord Lieutenant of County Durham, 1842," *Durham University Journal*, n.s., 38 (June 1978), 169–77.

"The Northern Coal-Owners and the Opposition to the Coal Mines Act of 1842," *International Review of Social History*, 25 (1980), 236–71.

"Debate: Coal, Class and Education in the North-East," *Past & Present*, no. 90 (February 1981), 136–42.

Hicks, J. R., "Annual Survey of Economic Theory: The Theory of Monopoly," *Econometrica*, 3 (1935), 1–20.

Hiskey, Christine E., "The Third Marquess of Londonderry and the Regulation of the Coal Trade: the Case Re-opened," *Durham University Journal*, n.s., 44, no. 2 (1983), 1–9.

Hopkins, Eric, "Religious Dissent in Black Country Industrial Villages in the First Half of the Nineteenth Century," *Journal of Ecclesiastical History*, 34, no. 3 (1983), 411–24.

Huberman, Michael M., "The Economic Origins of Paternalism: Lancashire Cotton Spinning in the First Half of the Nineteenth Century," *Social History*, 12, no. 2 (1987), 177–92.

"The Economic Origins of Paternalism: Reply to Rose, Taylor, and Winstanley," *Social History*, 14, no. 1 (1989), 99–103.

Hughes, Edward, "The First Steam Engines in the Durham Coalfield," *Archaeologia Aeliana*, 4th ser., 27 (1949), 29–45.

Jaffe, James A., "The State, Capital, and Workers' Control During the Industrial Revolution: The Rise and Fall of the North-East Pitmen's Union, 1831–2," *Journal of Social History*, 21: no. 4 (1988), 717–34.

"The 'Chiliasm of Despair' Reconsidered: Revivalism and Working-Class Agitation in County Durham," *Journal of British Studies*, 28: no. 1 (1989), 23–42.

"Competition and the Size of Firms in the North-East Coal Trade, 1800–1850," *Northern History*, 25 (1989), 235–55.

Joyce, Patrick, "Labour, Capital and Compromise: A Response to Richard Price," *Social History*, 9, no. 1 (1984), 67–76.

"Languages of Reciprocity and Conflict: A Further Response to Richard Price," *Social History*, 9, no. 2 (1984), 225–31.

Landsman, Ned, "Evangelists and their Hearers: Popular Interpretation of Revivalist Preaching in Eighteenth-Century Scotland," *Journal of British Studies*, 28, no. 2 (1989), 120–49.

Large, David, "The Election of John Bright as Member for Durham City in 1843," *Durham University Journal*, n.s., 16, no. 1 (1954), 17–23.

"The Third Marquess of Londonderry and the End of Regulation, 1844–5," *Durham University Journal*, n.s., 20, no. 1 (1958), 1–9.

Laslett, J. H. M., "The Independent Collier: Some Recent Studies of Nineteenth Century Coalmining Communities in Britain and the United States," *International Labor and Working Class History*, no. 21 (Spring 1982), 18–27.

Lazonick, William, "Industrial Relations and Technical Change: the Case of the Self-acting Mule," *Cambridge Journal of Economics*, 3 (1979), 231–62.

Lee, Charles, "The World's Oldest Railway: 300 Years of Coal Conveyance to the Tyne Staiths," *Transactions of the Newcomen Society*, 25 (1945–7), 141–62.

"The Wagonways of Tyneside," *Archaeologia Aeliana*, 29 (1959), 135–202.

Lee, D. J., "Skill, Craft and Class: A Theoretical Critique and a Critical Case," *Sociology*, 15, no. 1 (1981), 56–78.

Littler, Craig, R., and Salaman, Graeme, "Bravermania and Beyond: Recent Theories of the Labour Process," *Sociology*, 16, no. 4 (1982), 251–69.

Luker, David, "Revivalism in Theory and Practice: The Case of Cornish Methodism," *Journal of Ecclesiastical History*, 37, no. 4 (1986), 603–19.

Mann, Michael, "The Social Cohesion of Liberal Democracy," *American Sociological Review*, 35, no. 3 (1970), 423–39.

McCord, Norman, "The Government of Tyneside, 1800–1850," *Transactions of the Royal Historical Society*, 5th ser., 20 (1970), 5–30.

McCord, Norman, and Rowe, D. J., "Industrialization and Urban Growth in North-East England," *International Review of Social History*, 22 (1977), 30–64.

Metcalfe, Alan, "Organized Sport in the Mining Communities of South Northumberland, 1800–1889," *Victorian Studies*, 25, no. 4 (1982), 469–95.

Milne, Maurice, "Strikes and Strike-Breaking in North-East England, 1815–1844: The Attitude of the Local Press," *International Review of Social History*, 22 (1977), 226–40.

Milward, Neil, "Piecework Earnings and Workers' Control," *Human Relations*, 25, no. 4 (1972), 351–76.

Mitcalfe, W. Stanley, "The History of the Keelmen and their Strike in 1922," *Archaeologia Aeliana*, 4th ser., 14 (1937), 1–17.

Mokyr, Joel, "Demand vs. Supply in the Industrial Revolution," *Journal of Economic History*, 37, no. 4 (1977), 981–1008.

Mott, R. A., "English Wagonways of the Eighteenth Century," pts 1–2, *Transactions of the Newcomen Society*, 33 (1960–1), 1–33.

 "The Newcomen Engine in the Eighteenth Century," *Transactions of the Newcomen Society*, 35 (1962–3), 69–86.

Pallister, R., "Educational Investment by Industrialists in the Early Part of the Nineteenth Century in County Durham," *Durham University Journal*, n.s., 30 (1968–9), 32–8.

Pollard, Sidney, "Capital Accounting in the Industrial Revolution," *Yorkshire Bulletin of Economic and Social Research*, 15, no. 2 (1963), 75–91.

 "Fixed Capital in the Industrial Revolution," *Journal of Economic History*, 24, no. 3 (1964), 299–314.

 "A New Estimate of British Coal Production, 1750–1850," *Economic History Review*, 2nd ser., 33, no. 2 (1980), 212–35.

Porter, J. H., "Wage Bargaining under Conciliation Agreements, 1860–1914," *Economic History Review*, 2nd ser., 23, no. 3 (1970), 460–75.

 "Wage Determination by Selling Price Sliding Scales, 1870–1914," *Manchester School of Economic and Social Studies*, 39, no. 1 (1971), 13–21.

Price, Richard, "The Labour Process and Labour History," *Social History*, 8, no. 1 (1983), 57–75.

 "Conflict and Co-operation: A Reply to Patrick Joyce," *Social History*, 9, no. 2 (1984), 217–24.

Reid, Donald, "Industrial Paternalism: Discourse and Practice in Nineteenth-Century French Mining and Metallurgy," *Comparative Studies in Society and History*, 27, no. 4 (1985), 579–607.

Richards, E., "The Industrial Face of a Great Estate: Trentham and Lilleshall, 1780–1860," *Economic History Review*, 2nd ser., 27, no. 3 (1974), 414–30.

Richardson, R., "Primitive Methodism: Its Influence on the Working Class," *Primitive Methodist Quarterly Review* (1883), 261–73.

Rose, Mary, Taylor, Peter, and Winstanley, Michael J., "The Economic Origins of Paternalism: Some Objections," *Social History*, 14, no. 1 (1989), 89–98.

Rowe, D. J., "The Decline of the Tyneside Keelmen in the Nineteenth Century," *Northern History*, 4 (1969), 111–31.

"The Economy of the North-East in the Nineteenth Century: A Survey," *Northern History*, 6 (1971), 117–47.

"A Trade Union of the North-East Seamen in 1825," *Economic History Review*, 2nd ser., 25, no. 1 (1972), 81–98.

Rubery, Jill, "Structured Labour Markets, Worker Organisation and Low Pay," *Cambridge Journal of Economics*, 2, no. 1 (1978), 17–36.

Scott, Hylton, "The Miners' Bond in Northumberland and Durham," *Proceedings of the Society of Antiquaries of Newcastle-upon-Tyne*, 4th ser., 11: nos. 2, 3 (1947), 55–78; 87–98.

Sider, Gerald M., "The Ties that Bind: Culture and Agriculture, Property and Propriety in the Newfoundland Village Fishery," *Social History*, 5, no. 1 (1980), 1–39.

Sill, Michael, "Landownership and the Landscape: A Study of the Evolution of the Colliery Landscape of Hetton-le-Hole, Co. Durham," *Durham County Local History Society*, bulletin no. 23 (1979), 2–11.

"Mid-Nineteenth Century Labour Mobility: The Case of the Coal-Miners of Hetton-le-Hole, Co. Durham," *Local Population Studies*, 22 (1979), 44–50.

"Landownership and Industry: The East Durham Coalfield in the Nineteenth Century," *Northern History*, 20 (1984), 146–66.

Simpson, T. V., "Old Mining Records and Plans," *Transactions of the Institution of Mining Engineers*, 81 (1930–1), 75–108.

Smailes, Arthur E., "The Development of the Northumberland and Durham Coalfield," *Scottish Geographical Magazine*, 51, no. 4, (1935), 201–14.

"Population Changes in the Colliery Districts of Northumberland and Durham," *Geographical Journal*, 91 (1938), 220–32.

Spring, David, "The English Landed Estate in the Age of Coal and Iron: 1830–1880," *Journal of Economic History*, 11, no. 1 (1951), 3–24.

"The Earls of Durham and the Great Northern Coalfield, 1830–1880," *Canadian Historical Review*, 23, no. 3 (1952), 237–53.

"Agents to the Earls of Durham in the Nineteenth Century," *Durham University Journal*, 54, no. 3 (1962), 104–13.

Sraffa, Piero, "The Laws of Return under Competitive Conditions," *Economic Journal*, 36, no. 144 (1926), 535–50.

Stigant, P., "Wesleyan Methodism and Working-Class Radicalism in the North, 1792–1821," *Northern History*, 6 (1971), 98–116.

Swan, E. W., "A Durham Collieries' Stocktaking of 1784," *Transactions of the Newcomen Society*, 24 (1943–5), 109–112.

"Nicholas Wood's Ms. Report Book," *Transactions of the Newcomen Society*, 25 (1945–7), 139.

"Sinking a Northumberland Colliery in 1761–2," *Transactions of the Newcomen Society*, 25 (1945–7), 27–36.

Taylor, A. J., "Combination in the Mid-Nineteenth Century Coal Industry," *Transactions of the Royal Historical Society*, 5th ser., 3 (1953), 23–39.

"The Third Marquess of Londonderry and the North-Eastern Coal Trade," *Durham University Journal*, n.s., 17, no. 1 (1955), 21–7.

Thompson, E. P., "The Moral Economy of the English Crowd in the Eighteenth Century," *Past & Present*, no. 50 (February 1971), 76–136.

"Anthropology and the Discipline of Historical Context," *Midland History*, 1, no. 3 (1972), 41–55.

"On History, Sociology and Historical Relevance," *British Journal of Sociology*, 27, no. 3 (1976), 387–402.

Turner, E. R., "The Keelmen of Newcastle," *American Historical Review*, 21, no. 3 (1916), 543–5.

Ville, Simon, "Note: Size and Profitability of English Colliers in the Eighteenth Century–A Reappraisal," *Business History Review*, 58 (Spring 1984), 103–25.

"Total Factor Productivity in the English Shipping Industry: The North-East Coal Trade, 1700–1850," *Economic History Review*, 2nd ser., 39, no. 3 (1986), 355–70.

"Defending Productivity Growth in the English Coal Trade During the Eighteenth and Nineteenth Centuries," *Economic History Review*, 2nd ser., 40, no. 4 (1987), 597–602.

Walker, R. B., "The Growth of Wesleyan Methodism in Victorian England and Wales," *Journal of Ecclesiastical History*, 24, no. 3 (1973), 267–84.

Williams, Dale Edward, "Morals, Markets and the English Crowd in 1766," *Past & Present*, no. 104 (August 1984), 56–73.

Zeitlin, Jonathan, "Social Theory and the History of Work," *Social History*, 8, no. 3 (1983), 365–74.

"From Labour History to the History of Industrial Relations," *Economic History Review*, 2nd ser., 40, no. 2 (1987), 159–84.

Index

accommodationism, 97, 190–5
age-specific sex ratios, 79–81
Anglican Church, *see* Church of England
Antrim (Anne Catherine McDonnell),
    countess of, 55
arbitration; *see also* bargaining; piece rates
    and bond, 101–3
    after 1871, 193–4
Arkle, Robert, 70, 163, 177
asceticism, 121, 140
Ashton, T. S., 30
Atkinson, Robert, 67

Backhouse, Edward, 23–4
banks, 20, 23–5
Banks, Sir Edward, 25
bargaining, 2–3, 61–4, 71–2, 96–7, 105–11,
    194, 198, 199
    and bond, 111–12
    and working-class culture, 116–17
Bell, William, 33
bonds
    binding money, 75, 97
    fitters', 21–2
    miners', 62–3, 74, 100–4, 152–5
bord-and-pillar, 99
Bourne, Hugh, 131, 132, 134
Bouverie, Maj. Gen. Sir Henry, 184, 186–7,
    188
Brandling, R. W., 33–6, 91–2, 93, 156, 180,
    189
Braverman, Harry, 200
Brougham, Henry (Baron Brougham and
    Vaux), 93
Bryers, John, 97, 107, 108
Buddle, John, 14, 22, 33–6, 38–9, 41–2,
    43–6, 52, 54, 57, 58, 63–4, 69, 71, 75,
    76, 86, 100, 116, 118, 144, 146, 151,
    155, 158, 165, 167–8, 169, 173, 180,
    181, 184–5, 186, 187, 189
Bunting, Jabez, 137
Burawoy, Michael, 198–9, 200–1

Burgess, Keith, 191, 192
Burt, Thomas, 77, 85, 131, 140, 194

capital
    capital–output ratios, 15–16
    circulating, 19–25
    fixed, 10–19
casters, 41
cavilling, 84, 99, 117, 119
checkweighmen, 192–3
child labor, 82–7, 152–3
cholera, 145–7, 182
Church of England
    collieries, 51, 53–4, 127
    education, 88
    parish structure, 125–6
    patronage, 126–7
    pluralism, 124, 126
Church, Roy, 28
Clowes, William, 131, 132
coal industry
    coastwise markets of, 26–7
    and consumption, 26, 28
    export market of, 28
    output of, 27–8
    production costs of, 11–12, 15, 17–19,
    20, 37–40
    profits of, 38–9
coal miners, 61, 98; *see also* bargaining;
    Hepburn's union; piece rates
    autonomy of, 99–100
    as common and skilled laborers, 98, 104–
    5, 114
    and custom, 116–17
    and labor market, 73–6, 81–3
    and market ideology, 114–16, 199
    and piece work, 113–16
    and population movement, 76–9
coal owners' cartel, 22, 26, 30–6, 42–7, 62,
    115
    and Hepburn's union, 156, 164–5, 182–6
Cobbett, William, 125